W9-CPD-473

The Collected Papers of Albert Einstein

THE COLLECTED PAPERS OF ALBERT EINSTEIN

Volume 1

The Early Years: 1879 - 1902

Anna Beck, Translator

Peter Havas, Consultant

Princeton University Press

Princeton, New Jersey

Published by Princeton University Press
41 William Street
Princeton, New Jersey 08540

In the United Kingdom:
Princeton University Press
Chichester, West Sussex

ISBN: 0-691-08475-0 (paper)
0-691-08463-7 (microfiche)

Publication of this translation has been aided by a grant from the National Science Foundation

Printed in the United States of America by Princeton Academic Press

10 9 8 7 6 5 4 3

CONTENTS

List of Texts

PUBLISHER'S FOREWORD

We are pleased to be able to publish this companion volume to the documentary edition of *The Collected Papers of Albert Einstein*, Volume 1, thereby making available an English translation for those readers who cannot understand the original-language documents. This translation volume is intended to be only a supplement to the original edition, and therefore is available only to purchasers of the latter.

This publication project is separate from the documentary-edition project, and the editors of the documentary edition are not responsible for the accuracy of the translations. The documents were translated by Dr. Anna Beck; Dr. Peter Havas was our consultant. We are grateful to them for their hard work and dedication to this project.

We are also pleased to acknowledge a grant from the National Science Foundation, which made this publication possible. In particular, we thank Dr. Ronald Overman of the NSF for his continued interest in our work.

Princeton University Press

March 1987

PREFACE

The present volume provides a translation of all the documents contained in Volume 1 of *The Collected Papers of Albert Einstein*, all but two of which were originally written in German. The volume is not self-contained and should be read only in conjunction with the documentary edition. Most of the introductory matter, and all of the editorial headnotes, footnotes, and footnote numbers have been omitted, as have the figures, illustrations, biographical sketches, appendixes, bibliography, and index, most of which are already presented in English in the documentary edition.

The purpose of the translation, in accordance with the agreement between Princeton University Press and the National Science Foundation, is to provide "a careful, accurate translation that is as close to the German as possible while still producing readable English," rather than "a 'literary' translation." This type of translation should allow readers who are not fluent in German to make a scholarly evaluation of the content of the documents as well as obtain an appreciation of their flavor, in particular that of the correspondence. If some of the passages sound awkward, it is usually because the original passages were awkward - both because many of the letters and notes were obviously written in haste, and because the writers (especially Mileva Marić, whose native language was not German) did not always express themselves in correct, not to say literary, German.

Of note are some particular problems that arose in translating the correspondence. For example there is no way to render the distinction between the formal "Sie" and the informal "Du" (for "you") in modern English, and for this the original documents should be consulted. Similarly, for certain salutations there is no satisfactory English translation, and we have thus rendered "Sehr geehrter Herr Professor," for example, as "Esteemed Herr Professor." Whenever dialect expressions (as distinct from colloquial ones) were used - usually humorously - these are rendered in proper English, with the result, however, that the humor is often lost.

Both Einstein and Mileva Marić used diminutives to excess. Since there is no proper English equivalent for them, they are usually rendered as "little...", producing an awkwardness not present in the original, and hence are occasionally omitted in the translation. A similar problem arose concerning the numerous terms of endearment. We left "Dockerl" or "Doxerl" (little doll), a term frequently used by Einstein for Mileva, untranslated. "Miezchen" may or may not have its origin in "Miza," one of the Serbian diminutives for "Mileva" (which was never used by Einstein in the correspondence, however). Because it is also a standard South German expression for "pussycat" and may have been used by Einstein with this meaning, we thought it best to leave the term untranslated. Also, Einstein frequently referred to himself (and was addressed by Mileva) as "Johonzel" or "Johannzel" (with variations in spelling). The origin of this term, equivalent to "Johnny," is not known, and we left it untranslated. Misspelled names of persons and places (quite frequent, particularly in Einstein's letters and even in his scientific notes and papers) are routinely corrected without comment; other misspellings, of course, could not be maintained in translation.

Many technical expressions used in the original documents are outdated (see editorial notes in Volume 1). Wherever possible, we have not replaced them by the modern English ones; instead, we have used the expressions employed in the technical literature of the time, if known, or provided a literal translation. Similarly, "Israelit" and "israelitisch" are rendered as "Israelite," since these were terms used in Central Europe in the 19th and early 20th century.

We only rarely had to insert comments; these appear in square brackets []. Figures, graphs, and drawings had to be omitted for technical reasons; their approximate location, corresponding as closely as possible to that in the original, is indicated by "[Fig.]." All formulas were included in a form as similar to that in Volume 1 as possible with the word processor at our disposal, including all the errors in the original documents.

In the translation of the Weber notes (Document No. 37) the bracketed number at the top left hand corner of every 10th page indicates the equivalent page number of the German text of this document.

We are indebted to John Stachel, Editor, and Robert Schulmann, Associate Editor of the Einstein Project; and to Herbert S. Bailey, Jr., retired Director, and Alice Calaprice, Senior Editor, of Princeton University Press, for their help and encouragement.

Anna Beck
Peter Havas

March 1987

ALBERT EINSTEIN --
A BIOGRAPHICAL SKETCH
by
MAJA WINTELER-EINSTEIN
(EXCERPT)

THE FAMILY

Albert Einstein was born of German Israelite parents, and was thus originally a German citizen, as were all of his known ancestors.

The Einstein family is fairly widespread in southern Germany, especially in Württemberg and Bavaria, and since, as is well known, Israelites often marry more or less distant relatives, the Einsteins are related to most other Israelite families in southern Germany. Nothing more specific is known about Albert Einstein's more distant ancestors. Abraham Einstein, Albert's paternal grandfather, died in the prime of his life, and his grandson never knew him. He lived in Buchau on the Federsee, and is said to have enjoyed a great and widespread reputation as an intelligent and upright man. His wife Hindel, Albert's paternal grandmother, died during her grandson's early childhood. Her intellectual powers, it seems, were not particularly outstanding. His grandfather on his mother's side was Julius Derzbacher, who took the family name Koch. He was from Jebenhausen, where he practiced his trade as a baker, at first in modest circumstances. Later he lived together with his brother in Cannstatt, and together they managed to build a considerable fortune in the grain trade. The brothers and their families shared a single household under the same roof. Their wives shared the cooking, each taking charge of and responsibility for it in weekly turns. If such an arrangement is rather rare, and not only in Germany, theirs was all the more remarkable because it lasted for decades without the least friction. As his commercial abilities showed, Julius Koch possessed a distinctly practical intelligence and great energy. Theorizing was completely foreign to him. With wealth came a desire to be a patron of the arts, which he undertook, however, in a petty manner, and in accord with the principles of his trade, that is, spending as little as possible on it. As a result, he often ended up bying copies rather than authentic paintings. He once took in a poor artist he happened to meet on one of his walks for the purpose of laying the foundation of a future ancestral portrait gallery. This was the origin of a childhood portrait of Albert Einstein, still in the possession of the author. It is doubtful that the poor painter ever earned more than a free room and board under this arrangement. On the other hand, it was quite all right with grandfather Koch if technical skill, in this case a "likeness," took the place of genuine art. His wife, Albert's maternal grandmother, was Jette Bernheim. She had a quiet and solicitous nature, and was also clearheaded and methodical, as is apparent from surviving school essays. She handled the difficulties sometimes produced by grandfather Koch's choleric disposition with disarming humor. She was truly the soul of that odd household of the two brothers and their families.

Albert Einstein's father, Hermann, was born in 1847 in Buchau.

He entered the *Realschule* in Stuttgart at the age of fourteen and left with the so-called One-Year-Volunteer Certificate, the possession of which released the young German intelligentzia from the compulsory three years of military ervice. In reality, this arrangement gave preferential treatment to the "better classes" which could afford better schooling for their sons, and so keep them from rubbing shoulders with the sons of the common people. - Hermann Einstein, it seems, showed a marked inclination for mathematics, and would have liked to pursue studies in this or some related field. His father's means, however, with a large family to maintain and two daughters to provide for, were too limited to allow Hermann to pursue his inclination. As a result, he decided to become a merchant. Perhaps this very potential, left fallow in the father, developed all the more strongly in his son Albert. Hermann Einstein served an apprenticeship in Stuttgart and then became a partner in a cousin's business in Ulm.

The financial means brought to the marriage by his wife, and the progress of the business, might have allowed his young family not just a carefree but a very prosperous life. The future seemed secure, and there was such complete harmony of character between Hermann and his wife that the marriage would not only remain untroubled throughout their lives, but would also prove to be, at each turn of fate, the one thing that was firm and reliable. Had Hermann remained in Ulm, his son Albert would also have been granted a more carefree youth. But the family's external circumstances were to change in the course of time.

A younger brother of Hermann Einstein, named Jakob, who later exerted a certain intellectual influence on Albert while he was growing up, finished his studies in engineering and wanted to start a plumbing and electrical business in Munich. Since his own means were insufficient, he prevailed upon his brother Hermann to join in the venture, both personally as business manager and with a large investment. And so the family moved to Munich at the beginning of 1882, when Albert was barely two years old. Begun modestly, at a time when all the world was beginning to install electric lighting, the enterprise had good prospects. But Jakob Einstein's plans were more ambitious. His fertile and manifold ideas led him, among other things, to construct a dynamo of his own invention, which he wanted to produce on a large scale. That required a larger plant, and substantial funds to start operating it. The entire family, and especially Hermann's father-in-law Julius Koch, participated financially and made the new enterprise possible. It is hard to say just why it never really flourished. Whether because the highly imaginative Jakob Einstein dissipated his energies, or because, as an impetuous optimist he never understood how to deal with realities - in short, business affairs grew progressively worse. The fault may also have lain with Hermann Einstein, Albert's father, who, owing to his more contemplative nature, may have lacked the qualities required of a businessman on a grand scale. Hermann Einstein had a particularly pronounced way of trying to get to the bottom of something, by examining it from every side, before he could reach a decision. And since everything could always be looked at from a new point of view, that particularly enterpreneurial trait of being decisive at the right moment about the right matters was impaired. In addition, he was endowed with an unfailing goodness of heart, a well-meaning nature that could refuse nothing to anyone. So even though Jakob Einstein, constantly seeking novelty and change and unable to learn from any failure, was an over-eager and even stubborn optimist, his brother

Hermann gave in to him out of sheer good nature before he was himself able to reach decisions in his business deliberations.

This was demonstrated during a further change in the sphere of activity of these two very different brothers.

Business sales were insignificant in Germany, while showing great promise in Italy. The Italian representative of the firm then proposed moving the plant to Italy. Jakob Einstein was at once so taken with the idea that he was able to persuade Hermann Einstein to make the change, literally sweeping him along. The firm in Munich was liquidated. The lovely estate with the villa in which Albert Einstein had spent a happy childhood was sold to a building contractor, who immediately turned the handsome grounds into a construction site, cutting down the magnificent old trees and erecting an entire row of ugly apartment houses. Until the time of their move the children had to watch from the house as these witnesses to their most cherished memories were destroyed.

The plant was then transferred to Pavia; the family moved to Milan in 1894 and a year later to Pavia. The success of the enterprise was so meager, however, that by 1896 it had to be liquidated. Not only were the assets of Albert Einstein's mother lost at this time, but significant contributions from relatives as well. The family had hardly anything left. Their excessive confidence in the firm's Italian representative, who had been brought in as a partner, apparently contributed to this unfortunate turn of events.

At this point the two brothers, so dissimilar in nature, went their separate ways. Without prejudice, Jakob took a step which Hermann could not decide upon: he accepted a position as an engineer with a large company, and soon won trust and respect. In contrast, Albert Einstein's father could not bring himself to take the same step and relinquish his professional independence. In particular, he did not want to bring suffering on his wife, who would have had great difficulty accommodating herself to any lower standing in the social scale. Against the perceptive advice of his still quite young son, he founded a third electrical factory, in Milan. His cousin and brother-in-law from Hechingen, mentioned earlier, was persuaded to finance the enterprise, even though he had lost money in the earlier ventures. A capable former master mechanic, who had accompanied the two bothers from Munich, was made technical manager of this small factory. But at this time, there already existed larger, more financially powerful enterprises of this type in Italy, against which a small factory in rented space could not compete, particularly since constant financial problems limited the range of its activities. The firm being doomed to failure from the start, another crisis occurred within a few years, liquidation was necessary, and most of the invested capital was lost.

With money provided by relatives, Hermann Einstein then turned to installing power stations, supplying whole villages with lighting. This time, success seemed to be his. But the many worries, the constant feeling of personal dependence on someone else's money (how much more difficult this is to bear than the merely occupational dependence on one's employer!), all of these burdens had undermined his health, which until then had been robust. He quickly succumbed to a serious heart ailment and died in October 1902. His sad fate did not permit him even to suspect that two years later his son would lay the foundation of his future greatness and fame by solving an urgent problem in physics.

But Albert Einstein's mother was still able to enjoy her son's importance. A tall woman, in radiant health, her gray eyes gazed out at the world, often with a waggish twinkle. She possessed a sound native wit. Her feelings were seldom given free rein and, although accustomed to an opulent household, she adjusted - with difficulty, but with understanding - to her altered circumstances. Married at 17, she learned early about the realities of life and always maintained a certain practical sense, though basically she had a warm and caring nature. She was very fond of music and played the piano splendidly. Perseverance and patience were characteristic of her, as evidenced, for example, in her complicated and time-consuming needlework.

YOUTH

Albert Einstein was born in Ulm on 14 March 1879. At his birth his mother was shocked at the sight of the back of his head, which was extremely large and angular, and she feared she had given birth to a deformed child. But the doctor reassured her, and after a few weeks the shape of the skull was normal. The child, very heavy from the outset, was always quiet and required little care. He would play by himself for hours. His grandma, on first seeing him some time later, threw up her hands in surprise, and repeated over and over again: "Much too fat! Much too fat!". Otherwise, he developed slowly in childhood, and he had such difficulty with language that those around him feared he would never learn to speak. But this fear also proved unfounded. When the 2.5-year-old was told of the arrival of a little sister with whom he could play, he imagined a kind of toy, for at the sight of this new creature he asked, with great disappointment, "Yes, but where are its wheels?" The children of family and relatives often got together in his parents' garden in Munich. Albert refrained from joining their boisterous games, however, and occupied himself with quieter things. When he occasionally did take part, he was regarded as the obvious arbiter in all disputes. Since children usually retain a very keen and unspoiled instinct for the exercise of justice, the general recognition of his authority indicates that his ability to think objectively had developed early.

His early thoroughness in thinking was also reflected in a characteristic, if strange, habit. Every sentence he uttered, no matter how routine, he repeated to himself softly, moving his lips. This odd habit persisted until his seventh year.

At the age of five he received his first instruction at home from a woman teacher. Music lessons on the violin began at the same time. The usually calm small boy had inherited from grandfather Koch a tendency toward violent temper tantrums. At such moments his face would turn completely yellow, the tip of his nose snow-white, and he was no longer in control of himself. On one such occasion he grabbed a chair and struck at his teacher, who was so frightened that she ran away terrified and was never seen again. Another time he threw a large bowling ball at his little sister's head; a third time he used a child's hoe to knock a hole in her head. This should suffice to show that it takes a sound skull to be the sister of an intellectual. This violent temper disappeared during his early school years.

As is well known, in Germany one uses the polite form "Sie" for adults and for people who are not members of one's family, while "Du" is used only within the family, among children, and between close friends. There was thus something impertinent, but also something naive and humorous in little Albert's way of addresing his music teacher with "Du, Herr Schmied...."

Music was played often and well at home. Even though the fundamentals of the art were often difficult for the children and threatened to spoil it for the boy, because of his natural ability he soon developed a liking for music, which even led to artistic accomplishment. His musical ability seems to have come from the Koch branch of the family, the mathematical and logical from the Einstein side. Incidentally, it is not that uncommon, far apart as these two fields seem to lie, for mathematical and musical talent to be joined in one person.

The boy was trained early in self-reliance, in contrast to the customary European child-rearing method which consists of over-anxious tutelage. The 3- or 4-year-old was sent through the busiest streets of Munich; the first time he was shown the way, the second, unobtrusively observed. At intersections he conscientiously looked right, then left, and then crossed the road without any apprehension. Self-reliance was already ingrained in his character and manifested itself prominently on various occasions in his later life.

The boy entered the public primary school (*Volksschule*) at the age of seven. There he had a rather strict teacher whose methods included teaching children arithmetic, and especially the multiplication tables, with the help of whacks on the hands, so-caled "Tatzen" (knuckle raps); a style of teaching that was not unusual at the time, and that prepared the children early for their future role as citizens. His thinking process unerratic and thorough, the boy was considered only moderately talented precisely because he needed time to mull things over and didn't respond immediately with the reflex answer desired by the teacher. Nothing of his special aptitude for mathematics was noticeable at the time; he wasn't even good at arithmetic in the sense of being quick and accurate, though he was reliable and persevering. Also, he always confidently found the way to solve difficult word problems, even though he easily made errors in calculation.

At home, the rule that schoolwork must be finished before play could begin was strictly observed, and his parents accepted no excuses for breaking this commandment. Very typical of young Albert's abilities were the games he chose to play. He filled his leisure time by working on puzzles, doing fretsaw work, and erecting complicated structures with the well-known "Anker" building set, but his favorite was building many-storied houses of cards. Anyone who knows how much patience and precision is required to build card houses three or four stories high will be amazed that a boy not yet ten years old was able to build them as high as fourteen stories. Persistence and tenacity were obviously already part of his character and would become more and more prominent later on. The same trait that helped to keep his mother from tiring of the most tedious and complicated needlework manifested itself in her son first in his play and later in his scientific work. Many have brilliant insights in the course of life, original thoughts which nonetheless lead nowhere. Only persistence that does not rest until all that is unclear is eliminated and all difficulties are overcome allows an idea to take shape and be recognized as truly one of genius.

When Albert entered public school, his religious instruction, then compulsory in Bavaria, also had to begin. A liberal spirit, undogmatic in matters of religion, brought by both parents from their respective homes, prevailed within the family. There was no discussion of religious matters or rules. But since Albert was legally obliged to receive religious instruction, he was taught at home by a distant relative; as a result, a deep religious feeling was awakened in him. He heard about divine will and works pleasing to God, about a way of life pleasing to God - without these teachings having been integrated into a specific dogma. Nevertheless, he was so fervent in his religious feelings that, on his own, he observed religious prescriptions in every detail. For example, he ate no pork. This he did for reasons of conscience, not because his family has set such an example. He remained true to his self-chosen way of life for years. Later religious feeling gave way to philosophical thought, but absolutely strict loyalty to conscience remained a guiding principle. His later advocacy of Zionism and his activities on its behalf came from this impulse: less in accordance with and on the basis of Jewish teachings than from an inner sense of obligation toward those of his race for whom an independent working place for scholarly activity in the sciences should be created, where they would not be discriminated against as Jews.

Young Albert entered *Gymnasium* at the age of 8.5. In accord with the school's humanistic orientation, primary emphasis was placed on classical languages, Latin and later Greek, while mathematics and the natural sciences received less emphasis. The clear, rigorous logical structure of Latin suited his talents, but Greek and modern foreign languages were never his forte. His Greek professor, to whom he once submitted an especially poor paper, went so far in his anger to declare that nothing would ever become of him. And in fact Albert Einstein never did attain a professorship of Greek grammar.

In Gymnasium, the boy was supposed to begin the study of algebra and geometry at the age of 13. But by that time he already had a predilection for solving complicated problems in applied arithmetic, although the computational errors he made kept him from appearing particularly talented in the eyes of his teachers. Now he wanted to see what he could learn about these subjects in advance, during his vacation, and asked his parents to obtain the textbooks for him. Play and playmates were forgotten. He set to work on the theorems, not by taking their proofs from books, but rather by attempting to prove them for himself. For days on end he sat alone, immersed in the search for solution, not giving up before he found it. He often found proofs by ways that were different from those found in the books. Thus, during this one vacation of a few months, he independently worked his way through the entire prospective Gymnasium syllabus. Uncle Jakob, who as an engineer had a comprehensive mathematical education, reinforced Albert's zeal by posing difficult problems, not without good-natured expressions of doubt about his ability to solve them. Albert invariably found a correct proof; he even found an entirely original one for the Pythagorean theorem. When he got such results, the boy was overcome with great happiness, and was already then aware of the direction in which his talents were leading him.

At the same time the philosophical spirit began to stir in him. A poor Jewish medical student of Polish nationality, for whom the Jewish community had obtained free meals with the Einstein family, provided the impetus and thus repaid richly with intellectual stimulation what he received in material benefit. It was he who initiated the youth into the world of philosophical thought. He discussed with him all of the questions raised by the youth thirsting for knowledge and recommended the reading of books on natural philosophy (*Kraft und Stoff* [*Force and Matter*]by Büchner, *Kosmos* by Humboldt, the *Naturwissenschaftliche Volksbücher* [*Popular Books on Natural Sciences*] by Bernstein, among others). Moreover, despite the difference in their ages, he treated the boy as an equal and friend. Whereas Uncle Jakob's style of teasing skepticism about his abilities always spurred him on anew, and the teachers at the Gymnasium pedantically looked more for ready answers than for the ability to probe and reflect, the more insightful medical student offered young Albert far more. For he invested his whole person in examining everything that engaged the boy's interest. This occurred at that very crucial age when the child matures into a thinking person. His scientific interests were broadened as a result; he was no longer engrossed solely in mathematics, but had already begun to concern himself with the fundamental problems of the natural sciences in general. Music served as his only distraction. He could already play Mozart and Beethoven sonatas on the violin, accompanied by his mother on the piano. He would also sit down at the piano and, mainly in arpeggios full of tender feeling, constantly search for new harmonies and transitions of his own invention. And yet it is really incorrect to say that these musical reveries served as a distraction. Rather, they put him in a peaceful state of mind, which facilitated his reflection. For later on, when great problems preoccupied him, he often suddenly stood up and declared: "There, now I've got it." A solution had suddenly appeared to him.

When the family moved to Italy in 1894, the decision was made to leave Albert in Munich to finish gymnasium. This was done to ensure an uninterrupted course of studies, as well as because of the Italian language, which was foreign to the boy. He boarded with a family in Munich, while relatives and acquaintances made sure that he did not lack family contacts. In this period he sent only laconically phrased letters to Milan from which little could be ascertained about his life, yet this did not attract particular notice.

Actually, he was very uncomfortable in school. The style of teaching in most subjects was repugnant to him; moreover, his home room teacher did not seem very well disposed toward him. The military tone of the school, the systematic training in the worship of authority that was supposed to accustom pupils at an early age to military discipline, was also particularly unpleasant for the boy. He contemplated with dread that not-too-distant moment when he will have to don a soldier's uniform in order to fulfil his military obligations. Depressed and nervous, he searched for a way out. Hence, when the professor in charge of his class (the same one who had predicted that nothing good will ever come of him) scolded him on some occasion, he obtained a certificate from the family doctor, presented it to the school principal and abruptly left to join his parents in Milan. They were alarmed by his high-handed behavior, but he most adamantly declared that he would not return to Munich, and reassured them about his future by promising them most definitely that he would independently prepare himself for the entrance examination to the Zurich Polytechnical School (ETH) in autumn. This was a bold decision for a 16-year-old, and he actually carried it out. His parents resigned themselves to the new situation with grave misgivings, but were persuaded to do all they could to further the plan.

According to the German citizenship laws, a male citizen must not emigrate after his completed sixteenth year; otherwise, if he fails to report for military service, he is declared a deserter. For this reason the steps necessary for emigration were taken as quickly as possible, and young Albert was to be stateless until he was later naturalized in Switzerland. He diligently resumed his mathematical and scientific studies, and worked already then through nearly all of Violle's large textbook; in addition, he gained some practical experience in the family factory. His work habits were rather odd: even in a large, quite noisy group, he could withdraw to the sofa, take pen and paper in hand, set the inkstand precariously on the armrest, and lose himself so completely in a problem that the conversation of many voices stimulated rather than disturbed him. An indication of remarkable power of concentration.

A freer life and independent work made of the quiet, dreamy boy a happy, outgoing, universally liked young man. He also began to familiarize himself with classical German literature.

Though at first he was acquainted only with Milan and Pavia, Italy made a great impression on him even with this limitation. The way of life, the landscape, the art - everything attracted him, and later, from afar, became an object of longing. The hot summer of 1895 was spent in Ariolo on the Gotthard, where young Albert gained a fatherly friend in the Italian minister Luzzatti, who happened to be staying there.

He did so well at his autodidactic preparations that at the beginning of October 1895, at the age of only 16½, he passed the entrance examination to the Federal Polytechnical School in Zurich with the best outcome in mathematical and scientific subjects but inadequate results in linguistic and historical ones. Because of these gaps in his education and because of his youth, his parents were advised to have their son attend the final year of a Swiss secondary school, but with the prospect of certain admission the following year, despite the fact that he would still be fully six months below the presribed age (18 years).

So it was that Albert came to the Cantonal School (*Kantonsschule*) in Aarau, a small Swiss town whose schools had a deservedly high reputation and as a result were often attended by foreigners, even by some from overseas. He found welcome and understanding, and thus right away felt very much at home in the family of a teacher at the school, a scholar of literary and historical subjects. If the Munich Gymnasium had left him with a bias against secondary schools, this was thoroughly dispelled by the ways of the Aarau school. No traces of either a commanding tone or the cultivation of authority worship were to be found. Pupils were treated individually, more emphasis was placed on independent, sound thought than on punditry, and young people saw in the teacher not a figure of authority, but, alongside the scholar, a man of distinct personality. His time in Aarau was thus very instructive for him in many ways and one of the best periods of his life. His general education was enriched and, with the graduation certificate (Maturitätszeugniss) in his pocket, he was able to enter the Zurich Polytechnical School in the autumn of 1896.

Translated Texts

1 BIRTH CERTIFICATE

Ulm, 15 March 1879

No. *224*

Before the undersigned registrar appeared today,
in person *known,*
the merchant Herman Einstein
residing *in Ulm Bahnhofstrasse B No. 135*
of *Israelite* religion, and reported that to
Pauline Einstein née Koch
his wedded wife
of *Israelite* religion
who lives *with him*
in *Ulm in his residence*
on the *fourteenth of March* of the year
one thousand eight hundred *seventy-nine* *in the morning*
at *half-past eleven* o'clock a child of *male*
sex has been born, which was given
the first name *Albert*

Read, approved and *signed*
Hermann Einstein

Registrar
Hartmann

2. PAULINE EINSTEIN TO FANNY EINSTEIN

[Munich] 1 August 1886

Yesterday Albert got his grades, once again he was ranked first, he got a splendid report card...

3. COMMENT ON THE PROOF OF A THEOREM

[1891-1895]

3d Theorem. The cylinder surface is a surface that can be spread out or unrolled into a plane (surface développable).

* If one imagines that one places through each of several side lines [generators] of the cylinder surface (a) a plane that contains this and the next side line, (b) a plane that touches the surface, then one obtains two prismatic spaces, one that is inscribed within the cylinder surface and one that circumscribes it. The lateral surface of each of these prismatic spaces can be spread into a plane, inasmuch as one can imagine that each of the constituent planes rotates around one of its bordering edges until it falls onto the extension of the neighboring plane and then constitutes *one* single plane with it. Since the surface of the cylinder lies between the lateral surfaces of the prismatic spaces and constitutes the limit

that can be approximated as closely as desired by the lateral surfaces by increasing the number of planes from which they are formed, the same holds for the surface of the cylinder.

[Note in margin:] * The proof is pointless because as well as we can assume that the prismatic space can be unrolled, the same could be said about the cylinder!

4. TWO PHILOSOPHICAL COMMENTS

[1891–1895]

> *Leibnitz* applied this ad-infinitum continuing division of a finite quantity also to matter, in order to arrive in this way at its true components, and *Herbart* rightly says about this: "Even before one has done the first cut through the clump under consideration, there is apparent the infinite possibility that this same cut could be carried out in an infinite number of *different* ways. Herewith, actually, the whole infinite division is accomplished all at once; and one has arrived at the ultimate elements, to wit in thought, which has been the only thing that mattered. These ultimate elements cannot be matter" (because in that case one would have to repeat anew these countless divisions a *countless* number of times, which is nonsensical). "From this one ought to conclude at once, as Leibnitz has already done: It is not true that matter ultimately consists of other matter; its true components are simple (simple essences, substances, monads). And this is in conformity with truth." (Herbart's Metaphysik).

It is wrong to infer from the imperfection of our thinking that objects are imperfect.

[....]

> Whether one, along with *Leibnitz*, *Poisson*, *Herbart*, et al., seriously wants to take the infinitesimally small for a truly indivisible element, or one wants, along with others, to take it only for a useful fiction, so as thereby supposedly to eliminate all metaphysical difficulties, and conveniently and quickly introduce the calculus, is irrelevant for the calculus, for the one as much as the other leads to the goal.

[Note in left margin:] Sense? [Note in right margin:] ?!?

5. ON THE INVESTIGATION OF THE STATE OF THE ETHER IN A MAGNETIC FIELD

[Summer? 1895]

On the Investigation of the State of the Ether
in a Magnetic Field.

The following note is the first modest expression of a few simple thoughts on this difficult topic. It is with reluctance that I am compressing them into an essay that resembles more a program than a treatise. As I was completely lacking in materials that would have enabled me to delve into the subject more deeply than by merely meditating about it, I beg you not to interpret this circumstance as a mark of superficiality. May the indulgence of the sympathetic reader match the humble feelings with which I present these lines.

At its inception, an electric current sets the surrounding ether in a kind of momentary motion whose nature it has not yet been possible to determine with certainty. Despite the continuance of the cause of this motion, i.e., the electric current, the ether remains in a potential state and forms a magnetic field. That the magnetic field is a potential state is proved by the permanent magnet, for the law of conservation of energy precludes here the possibility of a state of motion. The motion of the ether produced by an electric current lasts until the acting motor forces have been compensated by equivalent passive forces originating from the deformation produced by the motion of the ether.

The marvelous experiments of Hertz most ingeniously elucidated the dynamic nature of these phenomena, the propagation in space, as well as the qualitative identity of these motions with light and heat. I believe that it would be of great importance for the understanding of the electromagnetic phenomena also to undertake a comprehensive experimental investigation of the potential states of the ether in magnetic fields of all kinds, or, in other words, to measure the elastic deformations and the acting deforming forces.

Any elastic change of the ether at any (free) point in some direction has to be ascertainable from the change undergone by the velocity of an ether wave at this point and in this direction. The wave velocity is proportional to the square root of the elastic forces serving the propagation and inversely proportional to the ether masses to be moved by these forces. Since the density changes produced by elastic deformations are usually insignificant, they probably might be neglected in this case too. One might therefore state with very good approximation that the square root of the ratio of the change in propagation velocity (wave length) is equal to the ratio of the change in the elastic force.

I would not dare to decide what kind of ether waves -- light, or else electrodynamic waves -- and what method of measuring the wave length would be most suitable for the examination of the magnetic field; basically, this would not matter anyway.

Provided a change of the wave length in the magnetic field is ascertainable in any direction, the first question to be solved experimentally could be whether it is only the component of the elastic state in the direction of wave propagation that exerts an effect on the propagation velocity, or whether such an effect is also exerted by the component perpendicular to the direction, since it is a priori clear that in a regular magnetic field, be it shaped like a cylinder or a pyramid, the elastic states are at any point completely homogeneous perpendicular to the direction of the lines of force and different in the direction of the lines of force. Hence, if polarized waves are allowed to penetrate perpendicularly to the direction of the lines of force, the direction of the plane of vibration would be of significance for the propagation velocity -- provided the elastic force component perpendicular to the propagation of a wave would indeed have an effect on the propagation velocity. This is probably not the case, even though the phenomenon of double refraction does seem to point to it.

Once the question of how the three components of elasticity affect the velocity of an ether wave has been solved, one could proceed with the investigation of the magnetic field. To properly understand the state of the ether in the magnetic field, one would have to distinguish three cases:

1. Lines of force that unite at the north pole in a pyramid-like fashion

2. Lines of force that unite at the south pole in a pyramid-like fashion.

3. Parallel lines of force.

In these cases one should investigate the propagation velocity of a wave in the direction of the lines of force and perpendicular to them. This will undoubtedly yield the elastic deformations along with the cause of their formation once sufficiently accurate instruments for the measurement of the wave length have been constructed.

The most interesting, and also most subtle, case would be the direct experimental investigation of the magnetic field formed around an electric current, because the exploration of the elastic state of the ether in this case would permit us a look into the enigmatic nature of electric current. The analogy would also permit us to draw sure conclusions about the state of the ether in the magnetic field surrounding the electric current, provided the previously mentioned investigations attain their ends.

Quantitative investigations on the absolute magnitudes of the density and the elastic force of the ether cannot begin, in my opinion, until there are available qualitative results bound to firm conceptions; I believe that I must say only one more thing. Should it turn out that the wave length is not proportional to $\sqrt{A + k}$, where A denotes the elastic ether forces a priori and hence is a constant that has to be found empirically, and k denotes the (variable) intensity of the magnetic field, which is, of course, proportional to the relevant elastic forces produced, then the reason for it should be sought in the density changes of the moving ether produced by the elastic deformation.

First of all, however, it has to be possible to prove that there does exist a passive resistance against the production of the magnetic field by the electric current, and that this [resistance] is proportional to the length of the current circuit and independent of the cross section and material of the conductor.

6. TO CAESAR KOCH

[Pavia, Summer 1895]

My dear uncle!

I am realy very glad that you are still interested in my humble doings despite the fact that we could not see each other for such a long time and that I am such a terribly lazy letter writer. All the same, I hesitated to send you this writing, because it deals with a very special topic; besides, it is rather naive and imperfect, as might be expected from such a young fellow like myself. I shall not be the least offended if you do not read the stuff at all; however, I hope that you will appreciate it as a timid attempt of mine to overcome my laziness regarding letter writing, which I inherited from both of my dear parents.

As you already know, I should now enter the Polytechnikum in Zurich. This matter encounters considerable difficulties because I should be at least two years older for it. We shall write you in the next letter about the outcome.

Give my love to the dear aunt and your cute little children. Your

Albert

7. ALBIN HERZOG TO GUSTAV MAIER

[Zurich] 25.IX.1895.

Mr. Gustav Maier, Zurich V.

In response to your inquiry of the 24th of this month, I wish to advise you as follows: According to my experience it is not advisable to withdraw a student from the institution in which he had begun his studies even if he is a so-called "child prodigy." In the case with which we are dealing my advice is to persuade the person in question to complete his entire course of studies in his present institution and pass the matura examinations. If you, or the relatives of the young man in question, do not share my opinion, I shall permit -- under exceptional dispensation of the age stipulation -- that he undergo an entrance examination in our institution. In this I proceed from the assumption that the rector of the educational establishment in question is going to confirm in writing and to the fullest extent your information regarding the talents and intellectual maturity of the candidate.

The director of the Federal Polytechnikum:

Herzog

Encl. Program.

8. ENTRANCE REPORT OF THE GEWERBESCHULE, AARGAU KANTONSSCHULE

[ca. 26 October 1895]

[...]
Entered in autumn:
[...]

Grade 3:

Einstein, Albert 14/III 1879 Gym. Munich

G[erman]	*3 - 2*	Ph[ysics]	*2*
I[talian]	*3*	Ch[emistry] Must do catch-up work	
F[rench]	has gr. gaps	[History]	*3 - 4*
G[eometry] d[escriptive]	*3*	[Natural history]	*3*
M[athematics]	*2*	[provisionally accepted]	

9. GUSTAV MAIER TO JOST WINTELER

Zurich V, 26 October 1895
Mittelstr. 12

Esteemed friend!

[...]

I do find it important, however, that Robert Koch be under your *most immediate* supervision. If there exists the slightest difference in this respect compared with the room at your neighbor's, i.e., if this somehow results in lesser control by you, then I would like to ask you to *swap* the rooms, and Albert Einstein, who is much more mature than his cousin and therefore less in need of supervision, will certainly make this little sacrifice for the sake of his cousin if I ask him nicely for it.

[...]

GustMaier

10. AARGAU KANTONSSCHULE RECORD

[26 October 1895 - 3 October 1896]

No.

Personal Records

for the pupil *Einstein, Albert* from *Ulm*
born on *14 March 1879* in
son of *Hermann Einstein, merchant in Pavia*
religion *Israelite*
boarding house *Prof. Dr. Winteler*
previous education in *Munich, Luitpoldt Gymnasium*
admitted on *26 Oct. 1895* to Grade 3 *Gew[erbeschule]* [tech. school]
left on from Grade
with a *Mat[ura]* certificate
for

Marks

(The first number denotes diligence, the second achievement)

Grade	3 Gewerbeschule				4 Gewerbeschule			
School year	1895/96				1896/97			
Quarter	I	II	III	IV	I	II	III	IV
Conduct			–	*good*	*good*			
Absences			–	*2*	*2*			
German			*good 2-3*	*2-3*	*4*			
Latin								
Greek								

Marks
(cont.)

French	*good* *3-4*	*2* *3-4*	*4-5* *3-2*
English			
Italian	*2-3*	*5(low)*	
Hebrew			
Religious instruction			
History	*1* *1-2*	*2*	*5*
Geography			
Arithmetic & Algebra Geometry	{ *1* }	*1*	*6*
Descriptive Geometry	*3*	*3-4* *2*	*4* *4-5*
Geodesy			
Natural History	*Bot. 2* *Min.2-3*	*Bot. 1-2* *Min. 2-3*	*5*
Physics	*1-2*	*1-2*	*6-5*
Chemistry	*1* *Rem[ark]*	*2* *3*	*5*
Chemical Practicum			
Technical Drawing	*2* *3*	*3*	*5* *5-4*
Artistic Drawing	*2* *3*		*5* *4*
Calligraphy			
Stenography			
Singing			*6* *5*
Music *Violin*	*1*	*1-2*	*5-6*
Gymnastics			
Military Instruction			

REMARKS

1895/96 III Qu[arter] Chemistry: In classes not yet to be judged. Has to continue with private lessons in French & Chemistry & Natural History.

1896. April Def[initely] promoted with protest in French.
96/97 I Quarter The protest in French remains in effect.

11. HERMANN EINSTEIN TO JOST WINTELER

Pavia, 29 October 95

Your Honor!

Permit me, esteemed Herr Professor, to express my deepest gratitude to you and your family for the exceedingly kind and friendly hospitality extended to my son Albert. He writes with delight about his stay there and already feels as comfortable as at home.

I also have high hopes regarding the many intellectual benefits his stay there is going to bring him; the stimulating conversations in your house will also be of special benefit to his knowledge.

Since Albert still lags far behind in the modern languages, I am taking the liberty of asking you to spur him to the utmost diligence in that direction and to arrange help through private lessons if necessary.

Finally, I would also like to ask you kindly to let me know your boarding terms soon.

In the meantime please accept the most respectful compliments, Your Honor, from your most devoted

Hermann Einstein

Kindest regards to your whole family from me, my wife, and my daughter.

12. MINUTES OF THE TEACHERS' CONFERENCE, AARGAU KANTONSSCHULE

Conference of 8 Nov. 1895

[...]

596 disp. Einstein, [grade] 3 Tr[ade School], required to receive private lessons in French, Natural History, and Chemistry, has been exempted from Singing and Gymnastics upon request, as an alien he is at this school grade also exempted from military instruction.

13. JOST WINTELER TO GUSTAV MAIER

Aarau, 21 Dec. 95

Dear Friend!

[...]

Our youngsters were recently in Zurich and, to my surprise, for all the visiting had no time for you. Otherwise everybody is doing fine, I think, both of them have come to the right place. [...]

J. Winteler

14. HERMANN EINSTEIN TO JOST WINTELER

Pavia, 30 Decemb. 95

Esteemed Herr Professor!

Your kind Christmas greetings made me and my family very happy, and I beg you to accept my heartfelt thanks for them and for the detailed report about Albert. It is a great relief to know that my son is under such loving care which is not only concerned with his physical well-being but also promotes his intellectual and inner life in such a noble fashion.

At this young age the heart is most receptive to a good model and I am convinced that your good influence will leave a lasting effect. I am of course exceedingly pleased with your opinion about Albert, even though I am aware that your words bespeak great benevolence.

I am taking the liberty of returning the enclosed school report; to be sure, not all its parts fulfil my wishes and expectations, but with Albert I got used a long time ago to finding not-so-good grades along with very good ones, and I am therefore not disconsolate about them.

Please allow me to send my best wishes for the New Year to you and your esteemed family on this occasion, and accept the most cordial greetings from your respectful and devoted

Hermann Einstein

15. PAULINE EINSTEIN TO THE WINTELER FAMILY

[Pavia, 30 December 1895]

Very esteemed family!

Permit me, too, to convey my best wishes for the New Year!

I feel so relieved and happy to know that my son is under such exquisite care and I wil be forever grateful for the great care with which you surround him.

Your little letter, dear Miss Marie, brought me immense joy, I will soon write to you, in the meantime my cordial greetings. We received your and Albert's note this morning.

Happy New Year!

Your utterly devoted

Pauline Einstein

16. RELEASE FROM WÜRTTEMBERG CITIZENSHIP

Ulm, *28 January 1896*

KINGDOM OF WÜRTTEMBERG
RELEASE CERTIFICATE

The undersigned Royal District Government hereby certifies that
Albert Einstein,
born on 14 March 1879 in Ulm
has been granted release from Württemberg citizenship upon his request

and in order to emigrate to *Italy*.

This Release Certificate brings about the loss of Württemberg citizenship for the persons expressly named herein, effective at the time of its delivery, but becomes invalid if the person released from citizenship does not transfer his domicile to an area outside the [German] Confederation or acquire citizenship in another Federal state within six months of the release certificate's delivery (§ 18 of the Law of 1 June 1870 on the Acquisition and Loss of Federal and State Citizenship; supplement of Regierungsblatt No.1, 1871, page 26)

Fee: 3 M[ark]
Tarif No. 69 No. 4

Royal Württemberg Government
for the Danube District
Hoser

Bühler

No.943

17. INSPECTOR'S REPORT ON A MUSIC EXAMINATION, AARGAU KANTONSSCHULE

[ca. 31 March 1896]

Report on the *instrumental music examination* at the Kantonsschule, which took place on 31 March of the current year from 2 to 4 o'clock.

Seventeen students were examined: 8 in piano and 9 in violin playing. [...]

The *violin playing* still revealed some stiffness in bowing techniques here and there, but otherwise the results were quite gratifying with regard to technique as well as with regard to intonation. One student, by name of Einstein, even sparkled by rendering an adagio from a Beethoven sonata with deep understanding. Hasler and Wohlwend also did very nicely.

Inspector:
J. Ryffel

vidit: Rödelberger
20 V 96

18. TO MARIE WINTELER, WITH A POSTSCRIPT BY PAULINE EINSTEIN

Pavia <Wednesday> Tuesday. Morning [21 April 1896]

Beloved sweetheart!

Many, many thanks sweetheart for your charming little letter, which made me endlessly happy. It is so wonderful to be able to press to one's heart such a bit of paper which two so dear little eyes have lovingly beheld and on which the dainty little hands have charmingly glided back and forth. I was now made to realize to the fullest extent, my little angel, the meaning of homesickness and pining. But love brings much happiness - much more so than pining brings pain. Only now do I realize how indispensible my dear little sunshine has become to my happiness. My mother has also taken you to her heart, even though she does not know you; I only let her read two of your charming little letters. Moreover, she always laughs at me because I am no longer attracted to the girls who were supposed to have enchanted me so much in the past. You mean more to my soul than the whole world did before, the "insignificant silly little sweetheart that knows nothing and understands nothing." Whether you should be afraid that ... (I am not going to repeat it)? If you were here at the moment, I would defy all reason and would give you a kiss for punishment and would have a good laugh at you as you deserve, sweet little angel! And as to whether I will be patient? What other choice do I have with my beloved, naughty little angel? The more so since the little angels are always weak (but indeed I am not going to say it again) & you are, after all, & should be, my little angel. How different it is to play a simple, sweet little song with one's sweetheart than to overcome, even if ever so "beautifully," an admittedly difficult sonata with straight as ramrod, decked-out Pavia ladies, with the ideal goal of "as fast as possible and at the same time as faultlessly as possible" hovering always in one's mind, or also "to extricate oneself from the situation with as much style as possible.["] The soul of the city can be roughly represented mathematically 1) from the sum of the ramrods swallowed by the various gentlemen and ladies 2) from the effect the everywhere uniform dirty walls & streets exert on the spirit of the spectator. The only beautiful thing are the charming, gracious little children; however, owing to the law of adaptation, the sight of the aforementioned repulsive things makes them become as insipid as their shining examples. Fortunately, my parents & my little sister are a very dear exception.

And now, I am sending you once again my heartfelt greetings my beloved child & be happy until the beautiful day of our reunion.

Your

Albert

Kindest regards to Mother & Manzis.

Without having read this letter, I send you cordial greetings!

Pauline Einstein

19. FINAL GRADES, AARGAU KANTONSSCHULE

Aarau, 5 Sept. 1896

FINAL GRADES OF THE STUDENTS OF THE 4TH TECHNICAL GRADE
(AUTUMN 1896)

Name and place of birth	*Born*	*Intends to study*
1. Einstein, Albert, Ulm	14/III 1879	Physics

German	*4 - 5*	*Geometry*	*6*
French	*3*	*Descr. Geometry*	*5*
*English**	*-*	*Physics*	*5 - 6*
*Italian**	*5*	*Chemistry*	*5*
History	*5*	*Natural History*	*5*
*Geography**	*4*	*Artistic Drawing**	*4*
Algebra	*6*	*Tech. Drawing**	*4*

[...]

* also matura grades

Dr. A. Tuchschmid
Rector

20. TO THE DEPARTMENT OF EDUCATION, CANTON OF AARGAU

Aarau, 7/IX/96

Department of Education of Canton Aargau.

I was born in Ulm on 14 March 1879 and at the age of one year got to Munich, where I lived until the winter of 1894-95. There I attended the elementary school and the Luitpold gymnasium up to (but not including) grade 7. Following that, until the fall of last year I lived in Milan, where I continued my studies privately. Since last autumn I have been attending the Kantonsschule in Aarau, and I am now taking the liberty of applying for the matura examination. After that I intend to study physics and mathematics at Department 6 of the Federal Polytechnikum.

Albert Einstein

21. *MATURA* EXAMINATION (A) GERMAN: "SYNOPSIS OF GOETHE'S GÖTZ VON BERLICHINGEN"

[18 September 1896, 7 - 9:20 A.M.]
Albert Einstein

SYNOPSIS OF GOETHE'S GÖTZ VON BERLICHINGEN

Two knights, Berlichingen and Weislingen, were intimate friends during their youth, but then due to the differences in their characters and inclinations, their paths of life take different directions. Götz von Berlichingen, a man of strong and independent character, is determined to assert his immediacy [exclusive dependence on the emperor and Reich, i.e., nondependence on the feudal princes] at all costs. From his castle he offers energetic resistance to the attacks of the Reich princes, especially of the Bishop of Bamberg.

In contrast, Weislingen, who is a talented and definitely good-natured man but with little strength of character, lives at the court of the Bishop of Bamberg, having exchanged the sweat and toil of freedom for the ease and comfort of the life at court.

Götz and Weislingen thus become open enemies. Götz succeeds in capturing the onetime friend of his youth. No sooner has he thus humiliated him than all the rancor he had harbored against him disappears. He gladly allows his sister Maria to become engaged to Weislingen, and only sets forth the condition that Weislingen move with her to his abandoned castle and avoid Bamberg.

The Bishop learns about this and is of course disconsolate at losing Weislingen. However, Liebetraut, a servant at the Bishop's court who is a sly fox of the highest order and knows well Weislingen's weaknesses, offers to bring Weislingen back by any means provided he'll be authorized to mix into his conversation something about Adelheid von Walldorf, a most beautiful widow visiting with the Bishop.

Liebetraut indeed succeeds in bringing Weislingen to Bamberg. Adelheid, an evil and ambitious person, knowingly lets herself be used as a tool of the Bishop. By means of her beauty and exceedingly sly behavior she succeeds in kindling such passion in Weislingen that he loses all control over himself. He breaks the promise he has made to Götz and leaves Maria, who later marries Sickingen, a powerful knight friendly with Götz.

Weislingen <goes to>, by lodging an accusation, succeeds in having the emperor authorize him to proceed against Götz, who now suddenly finds himself surrounded in his castle by a host of the emperor's troops. He is captured and brought to Heilbronn, where he is to swear an oath of eternal truce. The councillors of the city, which is hostile to him, demand much too far-reaching declarations so that they may use his refusal as a pretext to imprison him. From this exigency he is saved by Sickingen, who by force of arms marches into Heilbronn in the nick of time.

Scarcely has Götz peacefully settled in his castle for some time, when the Peasant War breaks out. Götz takes over the leadership, partly out of necessity and partly because he considered the uprising totally justified. He recognizes too late what a responsibility he has taken upon himself. The most terrible atrocities are taking place. He wants to detach himself from the insurgents. But it is too late. Weislingen has <suppressed> attacked the insurgent masses with his troops and has defeated them. Wounded in the battle and captured soon thereafter, Götz is condemned to death as a traitor by Weislingen.

In the meantime, the jealous Weislingen has ordered his wife to leave for his estates. Angry about that, and driven by her ambitious plans <Adelheit decided> she decides to poison Weislingen. She accomplishes this through his page, whose passionate love for her has made him her tool.

As he lies dying, Maria comes to plead with him for her brother's life. Under the most terrible physical and mental sufferings he signs <his> Götz's pardon and dies soon thereafter in despair.

Götz succumbs to his wound and with dark thoughts about the future takes leave of his wife, who had faithfully and firmly clung to him and revered him until his last breath.

Adelheid is condemned to death by the Vehmic court.

22. *MATURA* EXAMINATION (B) FRENCH: "MY FUTURE PLANS"

[18 September 1896, 2-4 P.M.]
Albert Einstein

MY PLANS FOR THE FUTURE

A happy man is too satisfied with the present to dwell too much upon the future. But on the other hand, young people especially like to contemplate bold projects. Also, it is natural for a serious young man to envision his desired goals with the greatest possible precision.

If I am lucky and successfully pass my examinations, I shall enrol in the polytechnical school in Zurich. I shall stay there four years to study mathematics and physics. I suppose I will become a teacher of these branches of natural science, opting for the theoretical part of these sciences.

Here are the reasons that have induced me to this plan. They are, most of all, my individual inclination for abstract and mathematical thinking, lack of imagination and of practical sense. My desires have also <brought forth the same goal> led me to the same <profession> decision. This is quite natural; everybody likes to do that for which he has a talent. Besides, I am also much attracted by a certain independence offered by the scientific profession.

23. *MATURA* EXAMINATION (C) GEOMETRY

[19 September 1896, 7-11 A.M.]
Albert Einstein

PROBLEM 1

Since the sides of a triangle are inversely proportional to the corresponding heights, we have

$$a = \frac{1}{h_1} \kappa = \frac{1}{2} \kappa$$

$$b = \frac{1}{h_2} \kappa = \frac{1}{3} \kappa$$

$$c = \frac{1}{h_3} \kappa = \frac{1}{4} \kappa$$

Since only the ratios of the sides are required for the determination of the angles, we pick the most useful one of the similar triangles, with sides 6, 4, and 3.

$$\cos \alpha = \frac{-a^2 + b^2 + c^2}{2\ bc} = \frac{-36 + 16 + 9}{24}$$

$$\cos \alpha = -\frac{11}{24}$$

$$\sin (\alpha - 90) = 0.4583$$

$$\lg \sin (\alpha - 90) = 9.66115 - 10$$

$$\alpha - 90 = 27°16'22''$$

$$\cos \beta = \frac{a^2 - b^2 + c^2}{2ac} = \frac{29}{36} = 0.8055$$

$$\log \cos \beta = 9.9061 - 10$$

$$\beta = 36°20'$$

$$\cos \gamma = \frac{a^2 + b^2 - c^2}{2ab} = \frac{43}{48}$$

$$\log \cos \gamma = 9.95226 - 10$$

$$\gamma = 26°23'$$

Calculation of side a

Since $\measuredangle\ \alpha$ is an obtuse angle,

$$a = 2r \cdot \sin\ (180° - \alpha)$$

$$\log a = \log 20 + \log \sin\ (64°43'38'')$$

$$= 1.30103 + 9.94884 - 10$$

$$= 1.24987$$

$$\underline{a = 17.77}$$

PROBLEM 2

If p denotes the distance between such a circle of the given system and its center, then its radius will be $= \sqrt{r^2 - p^2}$. Its equation is

$$(x - p)^2 + \langle r^2 - p^2 \rangle y^2 = r^2 - p^2$$

$$x^2 - 2px + p^2 + y^2 = r^2 - p^2$$

$$x^2 - 2px + y^2 = r^2 - 2p^2$$

We now search for the equation of the envelope, i.e., the intersection of two such circles whose p differ infinitesimally from each other. For the intersection, in the case of an infinitesimal increment $d(p)$, the increments of x and y as well as the equation must identically equal 0.

Hence: $$x^2 - 2px + y^2 - r^2 + 2p^2 = 0$$
$$\underline{x^2 - 2px + y^2 - r^2 + 2p^2 + (-2x + 4p)dp = 0}$$
Subtr. $$4p - 2x = 0.$$

We now substitute this value in the above equation

$$x^2 - 2px + y^2 - r^2 + 2p^2 = 0$$
$$x^2 - x^2 + y^2 - r^2 + \frac{1}{2}x^2 = 0$$
$$\frac{1}{2}x^2 + y^2 = r^2$$

For $x = 0$ $\quad y = \mp r$

For $y = 0$ $\quad x = \mp r\sqrt{2}$

We now have to consider the condition under which a circle of the system touches the ellipse $\frac{1}{2}x^2 + y^2 = r^2$.

<Since in the case of both figures (the circle from the system and the ellipse) the highest value for x has the ordinate 0 (as the 1st differential quotient shows), and the centers of all circles lie within the ellipse, we only have to examine whether for $y = 0$ one of the two points of the circle lies outside the ellipse. Since the entire figure under examination is symmetrical with respect to the y-axis, we need to consider only one (the positive) side.>

[Note in margin:] The proof is not sound.

We must make a direct comparison of the equations for the circle and for the ellipse and identify the x and y of both.

Ellipse $\qquad \frac{1}{2}x^2 + y^2 = r^2 \qquad$ I

Circle $\qquad x^2 - 2px + y^2 - r^2 + 2p^2 = 0$

<If we eliminate y:

$$\frac{1}{2}x^2 - 2px + \cancel{y^2} - \cancel{\frac{1}{2}x^2} - \cancel{r^2} + 2p^2 = 0$$

$$x^2 - 4px + 4p^2 = -4p^2 + 4p^2$$

The expression $\qquad \sqrt{-4p^2 + 4p^2} \quad >$

$$x = 2p$$

Substituted in I

$$2p^2 + y^2 = r^2$$

$$y = \mp \sqrt{r^2 - 2p^2}\,.$$

For the root to be real, we must have

$$r^2 > 2p^2$$

$$p\sqrt{2} < r \qquad\qquad p < \frac{r}{\sqrt{2}}$$

If $p \geq \frac{r}{\sqrt{2}}$, there is no more contact with the ellipse.

24. *MATURA* EXAMINATION (D) PHYSICS: "TANGENT GALVANOMETER AND GALVANOMETER"

[19 September 1896, 2:30–3:45 P.M.]
Albert Einstein

TANGENT GALVANOMETER AND GALVANOMETER.

Each electric current is surrounded by circular, concentric magnetic lines of force that lie in planes perpendicular to the current circuit. At any given point in the surroundings the magnetic force present is inversely proportional to the square of the distance from the (rectilinear) conductor, and directly proportional to the current in the conductor.

The latter proportionality <determines> is used for the relative determination of the current in a conductor <from the magnitude>, because in two different measurements the currents are proportional to the magnetic forces. This is done using the tangent galvanometer.

The setup of the latter is as follows: On a stand there is mounted a metal ring in a vertical plane which can rotate around a vertical axis. Around it <can> is fixed an insulated wire which can be connected with the current-producing apparatus by two conducting setscrews. In the middle of the ring there is suspended a magnetic needle on a very fine thread; the needle can move freely in the horizontal plane. Of course, one can also use magnets in forms other than a needle (e.g., a small mirror of magnetized steel such as is used for reading with a telescope and scale). Around the mobile magnet there are two mobile copper casings which serve for damping its oscillating motion.

[Fig.]

The apparatus is set up for use in such a way that the metal frame lies in the plane of the meridian. If a current is passed around the metal ring, then two forces act on each pole of the needle.

1) the horizontal component H of the terrestrial magnetic force in the direction of the meridian

2) perpendicularly to the latter, the magnetic force of the current circuit K, which is proportional to the current I, i.e., equals $I \cdot \kappa$ (a constant value for the instrument).

[Fig.]

The diagonal of the force parallelogram represents the magnitude and direction of the resulting magnetic force. Hence the needle will assume its direction. If we denote by φ the angle of deflection (the angle between the needle and the magnetic meridian), then it follows directly from the diagram

$$\text{tang}\ \varphi = \frac{\text{magnetic force of the current}}{\text{terrestrial magnetic force}} = \frac{I\kappa}{H}$$

A second measurement yields an analogous equation:

$$\text{tang}\ \varphi' = \ldots\ldots \qquad \frac{I'\kappa}{H}$$

Division yields $\dfrac{\text{tang}\ \varphi}{\text{tang}\ \varphi'} = \dfrac{I}{I'}$.

Thus the currents are proportional to the tangents of the angles of deflection.

The magnitudes of the magnetic forces of the earth and of the needle do not affect this kind of relative measurement (the magnitude of the needle's magnetic force has been taken to be equal to 1, because it does not affect the magnitude of deflection, since it would have only furnished a proportionality factor to both magnetic force components which then would have been cancelled by division).

The galvanometer differs from the tangent galvanometer only in its form, insofar as it uses many current coils instead of a single coil and partly different forms of damping. Of course, the relationship between the tangents of deflection and the magnitude of the current remains the same. The <relative> force of the deflecting current can be relatively increased not only by increasing the number of coils, but also by introducing a third magnetic force that counteracts the terrestrial magnetic force and thus partly neutralizes it.

The tangent galvanometer and galvanometer can be used for:

1) detecting currents

2) measuring currents, since they are proportional to the deflection tangents, but one current and its deflection must be known beforehand.

3) measuring electromotive forces, which is done by passing a shunt current through the apparatus whose total resistance is known, one gets then $\mathcal{E} = IW$.

4) measuring resistance by the Wheatstone bridge.

25. *MATURA* EXAMINATION (E) NATURAL HISTORY: "EVIDENCE OF THE EARLIER GLACIATION OF OUR COUNTRY"

[21 September 1896, 7-9 A.M.]

Albert Einstein

EVIDENCE OF THE EARLIER GLACIATION OF OUR COUNTRY

In order to derive conclusions about past glaciation on the basis of effects still persisting today, it is necessary to study the effects of present-day glaciers and to compare them with those remnants of earlier times.

For that purpose we have only to consider the effects of glaciers that involve rocks and formations because these are the only lasting ones. Two factors can be distinguished: the effect of the ice itself, and the effect of the glacier stream which is formed by the meltwater.

The ice rubs against the walls of the glacier valley, and there leaves behind rock striations in the direction of the glacier, i.e., in the direction of the valley floor. Of course, this activity is also connected with loosening of large and small rock fragments, which fall on the glacier and build a landform zone covered by debris along the edges.

firn mountain firn [Fig.]
glacier glacier
medial moraine lateral moraine

If two glaciers unite, the debris strips unite as well and then appear in the middle (medial moraines, as opposed to lateral moraines). If, as is usually the case, several glaciers unite, then a whole system of medial moraines arises, and these moraines turn into a uniform covering layer at the lowest part of the glacier as a consequence of the melting of the ice. The glacier now melts at the bottom and deposits this layer at the floor. If this keeps on occurring at the same spot for a long time, a row of hills arises from this debris, which takes the form of the glacier border, i.e., it closes off the valley in a horse-shoe fashion, with the open side directed upwards. Such rows of hills are called moraine hills. They consist of angular rocks of all sizes. However, if the glacier retreats slowly, it will distribute these blocks of rock in the area of retreat quite uniformly. This retreat region will also be <landform> tongue-shaped like the glacier itself. If the moraine strips in the lower part of the glacier do not mix but form separate bands, the deposited rocks will form interrupted lines (rows). Thus the moraine hills indicate a maximum or a halt in the expansion of a glacier, and erratic boulders a retrogression, but most of all they are proof of an earlier presence. Thus, these blocks consist of alpine rocks and hence in general differ from the kind of rock prevalent at their site, are angular and sharp-edged, form systems.

[Fig.] Schematic representation

erratic blocks

glacier
moraine
region of retreat

The glacier also acts upon the rocks at its bottom. Softer rocks get <finely> ground, harder ones get striated by friction and rounded

moraine direction of valley moraine [Fig.]

by water. They are then carried away by the glacial water and deposited when the pushing force of the water becomes too weak to carry them further. Such deposited rounded, striated rocks are often found where there now flows an eroding stream, which does not deposit rocks because the deposition takes place in a lake that is farther up. If this lake is determined by the structure of the mountains, then it is certain that the glacier has reached beyond the lake during such deposition periods. If more than one period of such depositions can be discerned (gravel terraces), as for example in Aarau, then this is proof that the glacier went beyond the lake on more than one occasion, giving rise to deposition, and retreated as many times, which implies an era of erosion for the lands lying <further> below the lake.

All these criteria are found in Switzerland. The region of erratic boulders extends as far as Germany. The sites of such boulders of the same kind of rocks form wide strips which narrow toward the mountains and lead to a mountain region in which these rocks crop out. The characteristic streaking (alpine line) can be found on valley walls where there are no more <traces of> glaciers. It is possible to distinguish between various moraine systems, which indicate periodic halts, which [the halts] are analogous in different glacier regions. Gravel terrace formation is found frequently and points to different glacializations.

26. *MATURA* EXAMINATION (F) ALGEBRA

[21 September 1896, 9:30-11:30 A.M.]
Albert Einstein

Given are the distances l, m, n between the center of an inscribed circle and the corners of the triangle. Find the radius ρ

of the inscribed circle $l = 1,\ m = \frac{1}{2},\ n = \frac{1}{3}$.

PROBLEM 1

$$\sin\frac{\alpha}{2} = \frac{\rho}{l} = \rho$$

$$\sin\frac{\beta}{2} = \frac{\rho}{m} = 2\rho$$

$$\sin\frac{\gamma}{2} = \frac{\rho}{n} = 3\rho$$

Since for each Δ in general:

$$\sin^2\frac{\alpha}{2} + \sin^2\frac{\beta}{2} + \sin^2\frac{\gamma}{2} + 2\sin\frac{\alpha}{2}\sin\frac{\beta}{2}\sin\frac{\gamma}{2} = 1,$$

after the substitution of the above values the equation will read:

$$<16>14\,\rho^2 + 12\,\rho^3 - 1 = 0 \qquad \rho = \frac{1}{x}$$

$$<\rho>\ <\frac{x^2}{16\rho} + >\frac{<16>14}{x^2} + \frac{12}{x^3} - 1 = 0$$

$$<16>14x + 12 - x^3 = 0$$

or:

$$x^3 - <16>\ 14x - 12 = 0$$

Now we have to apply Cardan's formula

$$x = \sqrt[3]{-\frac{q}{2} + \sqrt{\left(\frac{q}{2}\right)^2 + \left(\frac{p}{3}\right)^3}} + \sqrt[3]{-\frac{q}{2} - \sqrt{\left(\frac{q}{2}\right)^2 + \left(\frac{p}{3}\right)^3}},$$

where $p = -<16>14,\ q = -<12><14>12$.

The discriminant

$$\left\langle\sqrt{\left(\frac{q}{2}\right)^2 + \left(\frac{p}{3}\right)^3}\right\rangle\left(\frac{q}{2}\right)^2 + \left(\frac{p}{3}\right)^3$$

is negative, consequently its root is irrational.

Hence the trigonometric method should be applied. Then

$$\cos u = \frac{-\frac{q}{2}}{\sqrt{\left[-\frac{p}{3}\right]^3}} = \langle \frac{6}{\ } \rangle$$

$$\log(\cos u) = \log 6 + \frac{3}{2}\log 3 - \frac{3}{2}\log 14 =$$

$$= 0.7\langle 6\rangle 7185 + 0.71568 - 1.71919 =$$

$$= 9.77\langle 3\rangle 474 - 10$$

$$u = 53^\circ\ \langle 54\rangle 28'\ \langle 1\rangle 4''$$

The <other> 3 roots are

$$2\sqrt{-\frac{p}{3}}\cos\frac{u}{3}$$

$$2\sqrt{-\frac{p}{3}}\cos\left(\frac{u}{3} + 120^\circ\right)$$

$$2\sqrt{-\frac{p}{3}}\cos\left(\frac{u}{3} + 240^\circ\right)$$

Only positive roots are usable in the problem. Since $2\sqrt{-\frac{p}{3}}$ is positive, the other factor must also be positive if the product is to be positive.

$\frac{u}{3}$ is an acute angle, hence its cosine positive, and the first root is usable. The cosine of the second angle (situated in the second quadrant) is negative, thus the root is unusable. The cosine of the third $\frac{u}{3} + 240^\circ$ is smaller than 260°, hence it still lies in the third quadrant, and hence the 3d root is unusable.

$$\log x = \log 2 + \frac{1}{2}\log\frac{14}{3} + \log\cos 17^\circ\ 49'\ 21''$$

$$\log x = 0.61420 \qquad \log 10\,\rho = 0.38580$$
$$\rho = 0.243$$

27. *MATURA* EXAMINATION (G) CHEMISTRY

[21 September 1896, 2-4 P.M.]
Albert Einstein

PROBLEM.

How many liters of 30% hydrochloric acid, whose specific gravity is 1.15, are obtained from 39.5 kg table salt when neutral sodium sulfate forms as the reaction side product. Describe the substances formed during the reaction.

For sodium sulfate and hydrochloric acid to be formed from table salt, table salt must react with sulfuric acid. This reaction occurs according to the following equation:

$$H_2SO_4 + 2\ NaCl = Na_2SO_4 + 2\ HCl.$$

The atomic weight of Na = 23, Cl = 35.5, H = 1, hence

117 parts by weight of table salt are equivalent to 73 parts by weight of HCl.

If 39.5 kg table salt are used and the weight of the HCl obtained is denoted by x, then we have the following proportion:

$$117:73 = 39.5:x$$

$$x = \frac{73 \cdot 39.5}{117} = 24.64 \text{ kg gaseous hydrochloric acid.}$$

However, since 30% hydrochloric acid has been used, we have for the weight of the diluted hydrochloric acid y the following equation:

$$30:100 = 24.64:y \qquad y = 82.1 \text{ kg}$$

If we denote the unknown volume by z, and since the specific gravity of 30% hydrochloric acid is 1.15, we have

$$1:1.15 = z:82.1$$

$$z = \frac{82.1}{1.15} = 71.4\ \ell.$$

Hence 71.4 ℓ 30% hydrochloric acid are formed.

Substances formed:

1) Na_2SO_4 Glauber's salt. Crystallizes with 10 molecules of water of crystallization, is water-soluble like all alkaline salts excepting the silicates, has a bitter taste, and is used in medicine as a laxative; it is found in natural mineral waters. It is an intermediate product in the soda process, where it is reduced to Na_2S (sodium sulfide) by calcination with carbon.
2) HCl hydrochloric acid. This is a colorless, pungent gas which easily dissolves in water. Between the weight and percent concentration of the solution there exists a relationship, which has been collected in tables, i.e., the specific gravity increases with increased HCl content. This is a rather strong acid, and it vigorously attacks metals and makes the litmus turn red. When brought together with ammonia, <it> the <gaseous> hydrochloric acid forms a precipitate NH_4Cl. Hydroiodic, hydrobromic and hydrofluoric acid are its analogs.

28. ETH RECORD AND GRADE TRANSCRIPT

[5-10 October 1896 - 2 August 1900]

REGISTER

Einstein, Albert, from Ulm, born *14 March, 1879.*
Address of parents or guardian: *Hermann Einstein, Via Foscolo 11, Pavia.*

DEPARTMENT VIA.

1st year curriculum	School year *1896/97.*
2d " "	" " *1897/98.*
3d " "	" " *1898/99.*
4th " "	" " *1899/1900.*

DEPOSITED DOCUMENTS:

Certificate of citizenship.
Birth certificate. Certificate of good conduct.
Matura certificate of the Gewerbeschule Aarau.

WITHOUT ENTRANCE EXAMINATION.

Composition	Mathematics
Political and literary history	Descriptive geometry
German language	Chemistry
French language	Physics
Natural sciences	Drawing

Admitted: *October 1896.*
Left: *August 1900.*

Remark: 6 is the best, 1 the lowest grade.

VIA DEPARTMENT

1st year curriculum 1896-1897.

Subjects	Instructor	I. Semester Performance	II. Semester Performance	Remarks
Differential & integral calculus with exercises	*Hurwitz*	4½	5	
Analytical geometry	*Geiser*	5		*Graduated.*
Descriptive geometry w. exercises	*Fiedler*	4½	4	
Mechanics w· exercises	*Herzog*		5	
Determinants	*Geiser*		–	
Projective geometry	*Fiedler*		4½	

Nonobligatory subjects.

1st semester		2d semester	
Prehistory of man:	*Heim.*	*Exterior ballistics:*	*Geiser.*
Central projection:	*Fiedler.*	*Kant's philosophy:*	*Stadler.*
		Goethe (Works & world view)	*Saitschick.*

VIA DEPARTMENT

2d year curriculum *1897-1898*.

Subjects	Instructor	I. Semester Performance	II. Semester Performance	Remarks
Differential equations	*Hurwitz*	5		
Physics	*Weber*	5½	5	*Graduated*
Mechanics	*Herzog*	5½		
Projective geometry	*Fiedler*	4		
Infinitesimal geometry	*Geiser*	-	-	
Geometry of numbers	*Minkowski*	-		
Number theory	*Rudio*	-		
Theory of scientific thinking	*Stadler*	-		
Geometric theory of invariants	*Geiser*		-	
Theory of functions	*Minkowski*		-	
Potential theory	*Minkowski*		-	
Theory of definite integrals	*Hirsch*		-	
Introduction to celestial physics	*Wolfer*		-	

Nonobligatory subjects

1st semester	2d semester	
	Banking & stock exchange:	*Platter.*
	Mathematical foundations of statistics & personal insurance:	*Rebstein.*
	Geology of mountains:	*Heim.*
	Politics of Switzerland:	*Öchsli.*

VIA Department
3d year curriculum 189*8*-189*9*.

Subjects	Instructor	I. Semester Performance	II. Semester Performance	Remarks
Elliptic functions	*Minkowski*	-		*March 1899: director's reprimand for nondiligence in physics practicum.*
Analytical mechanics	*Minkowski*	-		
Theory of linear differential equations	*Hirsch*	-		
Introduction to astronomy	*Wolfer*	-		
Celestial mechanics	*Wolfer*			*Graduated*
Electrotechnical principles, apparatuses & meth. of measurem.	*Weber*	-	-	
Electrical oscillations	*Weber*	-		
Electrotechnical laboratory	*Weber*	6	-	
Introduction to phys. practicum	*Pernet*		-	
Phys. pract. for beginners	*Pernet*	1		
Calculus of variations	*Minkowski*		-	
Algebra	"		-	
Scient. projects in phys. lab.	*Weber*		5	
Introduction to electromechanics	*Weber*		-	
Linear differential equations	*Hirsch*		-	
Alternating currents	*Weber*		-	
Determ. of geographic location	*Wolfer*		4½	
Exercises in astronomy	*Wolfer*		-	

<u>Nonobligatory subjects.</u>

1st semester

2d semester
Income distribution and the social consequences of free competition: Platter.

VIA DEPARTMENT

4th year curriculum 1899-*1900*.

Subjects	Instructor	I. Semester Performance	II. Semester Performance	Remarks
Partial Differential equations	*Minkowski*	-		
Alternating current systems & alternating current motors:	*Weber*	-		
System of absolute electrical measurements:	*Weber*	-		
Scient.projects in the physical laboratories	*Weber*	*6*	*5*	
Introduction to the theory of alternating currents	*Weber*		-	
Applications of analyt. mechanics	*Minkowski*		-	

Nonobligatory subjects.

1st semester		2d semester	
Foundations of national economy:	*Platter*	*Cultural history of Switzerland in the Middle ages and the Reformation Period*	*Oechsli.*

LEAVING CERTIFICATE

Subjects	Instructor	Performance
Differential & integral calculus with exercises	*Hurwitz*	*4 3/4*
Differential equations w. exercises	"	*5*
Descriptive geometry w. exercises	*Fiedler*	*4¼*
Projective geometry	"	*4¼*
Mechanics w. exercises	*Herzog*	*5¼*
Analytical Geometry	*Geiser*	*5*
Determinants	"	*5*
Infinitesimal geometry	"	–
Geometric theory of invariants	"	–
Geometry of numbers	*Minkowski*	–
Theory of functions	"	–
Potential theory	"	–
Elliptic functions	"	–
Analitical mechanics	"	–
Calculus of variations	"	–
Algebra	"	–
Partial differential equations	"	–
Applicat. of analytical mechanics	"	–
Number theory	*Rudio*	–
Theory of definite integrals	*Hirsch*	–
" " *linear differential equat.*	"	–
Introduction to physic. experiments	*Pernet*	–
Physical experiments for beginners	"	*1*
Physics	*Weber*	*5¼*
Electrotech. principles, apparatuses & methods of measurement	"	–
Electric oscillations	"	–
Electrotechnical laboratory	"	*6*
Scientific projects in phys. lab.	"	*5 1/3*
Introduction to electromechanics.	"	–
Alternating currents	"	–
System of absolute elect. measurements	"	–
Intro. to celestial physics	*Wolfer*	–
Celestial mechanics	"	–
Determination of geographic location	"	*4½*
Exercises in astronomy	"	–
Theory of scientific thinking	*Stadler*	–
Kant's philosophy	"	–
Nonobligatory subjects:		
Central projection	*Fiedler*	

Exterior ballistics	*Geiser*
Prehistory of man	*Heim*
Geology of mountains	"
Politics of Switzerland	*Oechsli*
Cultural history of Switz. in the Middle Ages & Reformation Period	"
Banking & stock exchange	*Platter*
Social consequences of free competition	"
Foundations of national economy	"
Mathem. foundations of statistics & personal insurance	*Rebstein*
Goethe: Works & world view	*Saitschick*

Remarks: *There are no complaints about moral conduct.*

Issued on *2 August 1900*

Diploma as: *Teacher of mathematics*
in accordance with the decision
of the Swiss School Council
or the order of the President
dated *28 July 1900.*

29. FROM MARIE WINTELER

Olsberg, Wednesday evening [4-25 November 1896]

Beloved sweetheart!

Your little basket arrived today and in vain did I strain my eyes looking for a little note, even though the mere sight of your dear handwriting in the address was enough to make me happy. I really thank you, Albert, for wanting to come to Aarau, and I don't have to tell you that I will almost be counting the minutes until that time. Tomorrow is Thursday and after that is Friday and then finally finally Saturday and then you will come with your fiddle, your dear child, and your other child (also dear?) will come from the other side. Who is going to come first and be able to pick up the other one at the railway station? I'll be home around 2:30 or 3 o'clock and don't you come so late either, sweetheart, I have to leave on Sunday before 4 o'clock, it is quite far to walk to Olsberg almost 1½ hours away from the railway station. Last Sunday I was crossing the woods in pouring rain to take your little basket to the post office, did it arrive soon? My love, I do not quite understand a passage in your letter. You write that you do not want to correspond with me any longer, but why not, sweetheart? You said already in your Toggenburg letter that we would want to write each other again when I'll be in Olsberg. However, I am complaining, am I not, but I would rather not do so and wait with it until [we] are together again. But then, I am not going to forget it, and if I'll get a really nice answer, such a one that I would really like and then we'll play together and...oh, I am so terribly glad. You scold me rudely that I don't want to write to you how and why I've come here. But you dear wicked one, don't you know that there exist a lot of more beautiful and more clever things one can chatter and tell about than something so stupid. And when you, great dear philosopher (but a totally false one) draw such wicked and not at all true conclusions (even the thoughts of such a clever darling curlyhead are not always logical, isn't that so?) as you have done in your last letter to your sweetheart, then I cannot grasp it at all and you must be quite annoyed with me if you can write so rudely. But wait you'll get some proper scolding when I come home, one learns that up here, sweetheart!

I am writing a lot of rubbish, isn't that so, and in the end you'll not even read it to the finish (but I don't believe that). But isn't it so, sweetheart, you know well all that dwells and lives in my heart for you and you alone, and I could never describe, because there are no words for it, how blissful I feel ever since the dear soul of yours has come to live and weave in my soul, all I can say is that I love you for all eternity, sweetheart, and may God preserve and protect you.

With deepest love yours

Mariechen

30. FROM MARIE WINTELER

Olsberg, 30 November 1896

My dear dear sweetheart!

Finally finally I felt happy happy, something only your dear dear letters can bring about, and your little note made me also completely healthy again. But I had to wait terribly long and had also written to mama to write to me whether maybe my sweetheart is ill. As far as I am concerned, you don't have to worry, darling, I am quite healthy and merry once again, and I can wait out the 12 days (because there are no 14 days left anymore) after which....oh how delightful! I do not think about myself, sweetheart, that's quite true, but the only reason for this is that I do not think at all, except when it comes to some tremendously stupid calculation that requires, for a change, that I know more than my pupils, who are proper little blockheads. And all this comes from the great bliss in the heart that makes one quite frivolous, isn't it so, sweetheart, are you now satisfied with me?

That you do not want to give me an answer, just you wait, Albert, you'll get quite a wicked punishment for it, I have still 12 days left in which to devise one (and all the same I am so glad that you don't want to give me one and that you think the whole thing stupid, isn't it curious that I like this even much better than an answer, but, then again, I would like one all the same, I don't know why).

My dear sweetheart, the "matter" of my sending you the stupid little teapot does not have to please you at all as long as you are going to brew some good tea in it, and then, sweetheart, you will get only what my heart can give to you, and nothing else, isn't that so. Now, be satisfied and stop making that angry face which looked at me from all the sides and corners of the writing paper.

Don't work too much, you dear. How I sometimes yearn to stroke a little your dear tired brow when I picture to myself how you now certainly sit in your little room tired and pensive, the way you often did back home, you know. In such moments I would like to fly to my sweetheart and tell him how much do I love you, and banish all the worries about work for a little while. My dear dear darling! After the sad times of my banishment up here have come to an end, mama and I will come once to Zurich to see where my darling dreams away his days, I am looking forward to that so much. Then I will arrange everything the way I like it, and you will enjoy your little study room twice as much.

And now some more "school talk." I like to teach very much, all the children love me, and I love them too. But obey they must sweetheart, and that only because I am so terribly cross. They are always as quiet as little mice, and they open their little mouths and noses so that a lot of wisdom (especially mathematics) may come in, and they are terribly respectful. A little boy in the first grade, who shares with you a facial feature and, imagine that, whose name is also Albert, has it especially good with me, I want to help him with his lessons and, in general, love him ever so much. It's sometimes quite strange, something comes over me when he looks at me and I always believe that you are looking at your little sweetheart. Good night, darling.

Thousands of heartfelt greetings from your

Mariechen

31. PAULINE EINSTEIN TO MARIE WINTELER

Milan, Sunday 13 Dec.96

My dear Miss!

[...]

The beautiful Christmas time now draws near again, to you it will be doubly welcome this year as it will reunite you with your family after whom you surely yearn very much. To us, too, it brings a most anxiously awaited visit, Albert will probably arrive one week from today. The rascal has become frightfully lazy, one can notice the absence of loving admonition, for the regularity of his letters leaves much to be desired. E.g., I have been waiting in vain for news for these last three days; I will have to give him a thorough talking-to once he'll be here. Will it help? [...]

Pauline Einstein

32. PAULINE EINSTEIN TO MARIE WINTELER

[Milan] Wednesday 24 March 1897

My dear Miss!

Each day I make plans to write you, but when Albert is here I do not accomplish anything: There is so much laughing, joking, and music-making going on that there is not enough time for anything else. A completely different life moved in with him; he looks excellent, he has even grown some more & he has developed a gigantic appetite. [...]

Maya has now taken over the role of her brother's accompanist, he is very satisfied with her, for she has improved very much. On the whole, the two are very happy with each other, sometimes their

boisterousness gets out of hand to such an extent that it drives me out of the room, but then I become even more the target of the two imps' jokes. Yes, maternal authority often gets totally subverted!!
[...]

Pauline Einstein

33. STATEMENT OF A FINE

Zurich, *23/28 April* 1897

IMPOSITION OF FINE
No. *6619*

As it has turned out *that Albert Einstein, stud. math., born 1879, from Ulm, Württbg., residing at Hägi's, Unionstrasse 4, District V, has been staying in Zurich since 28 January 1897 without having delivered valid identification documents,*
therefore, due to this violation of Art. 4 of the Order of the City Council concerning the delivery of documents and administration of the control of residents of 30 May 1894, a fine of *10* fr. is imposed on *Albert Einstein*.

This ruling cannot be appealed. However, within 10 days, counted from the notification on the decision, the fined person can request a court adjudication of the matter, which has to be done with the date and signature on the verso of this Order. Failure to respond will be taken as acceptance of the fine. (§1055 of the Zurich Regulations on Criminal Procedure).

The fine must be expeditiously paid to the cashier's office of the Central Control Bureau. After 14 days have passed without response, legal proceedings will be instituted, and the fine might be converted to imprisonment according to §1060 of the Zurich Law of Criminal Justice.

If valid identification papers are not deposited within a further 10 days, deportation by the police will ensue.

Chief of the Central Control Bureau:
Bühler

34. TO PAULINE WINTELER

Zurich, Thursday [May? 1897]

Dear mommy!

I am writing you so soon in order to cut short an inner struggle whose outcome is, in fact, already firmly settled in my mind: I cannot come to visit you at Whitsuntide. It would be more than unworthy of me to buy a few days of bliss at the cost of new pain, of which I have already caused much too much to the dear child through my fault. It fills me with a peculiar kind of satisfaction that now I myself have to taste some of the pain that I brought upon the dear girl through my thoughtlessness and ignorance of her delicate nature. Strenuous intellectual work and looking at God's Nature are the reconciling, fortifying, yet relentlessly strict angels that shall lead me through all of life's troubles. If only I were able to give some of this to the good child! And yet, what a peculiar way this is to weather the storms of life -- in many a lucid moment I appear to myself as an ostrich who buries his head in the desert sand so as not to perceive the danger. One creates a small little world for oneself, and as lamentably insignificant as it may be in comparison with the perpetualy changing size of real existence, one feels miraculously great and important, just like a mole in his self-dug hole. -- But why denigrate oneself, others take care of that when necessary, therefore let's stop.

Your dear little letter, the lilies of the valley, the little poems, all of them brought me great joy, like everything that comes from your dear little house. I thank you from all my heart for it. There is very little that is of interest in my external life: in fact, the latter is so philistine that people could use it for setting their watches -- except that their watches would be somewhat late in the morning. As for my intellectual life, there is always quite a variety. Saturday evenings I play music at the home of a local lady with a few other gentlemen, including Byland; these are the most beautiful hours of my week. Byland read to me a few plays by Gerhart Hauptmann, and these affected me tremendously. "Hanneles Himmelfahrt" made me cry like a child, half in bliss and half in pain. You too should read this gem; I cannot say more about it -- one must keep silent when one thinks about it.

Thousand greetings to you and your family from your

Albert

35. TO PAULINE WINTELER

Zurich, Monday [7 June 1897]

Dear mommy!

Your lovely present gives me a welcome excuse to write to you again, the holiday's silence, the cozy quietude, to have a good chat with you, as if we were sitting together in the red room while the potatoes are getting brown with jealousy and the dear sun and some other dear thing peep into the room. When I think of that room, my head starts ringing in a delightfully mad way, and a thousand memories, some old, some young, some gay and others sad, embrace each other in a child-like fashion, as if they belonged together. Many an old philosophical deduction in a long house robe with unmended holes paces there solemnly in the air, and next to it giggles many a charming and foolishly sweet little word with little wings and rosy cheeks -- and thank God, they are far more numerous, and, sweetly making fun of me, they still grab me sometimes by the nose when I, with knitted brow, cultivate the golden scholarship in my room. And afterwards I feel so silly, and curiously vacillating between laughter and tears -- and finally the beloved piano resounds like my soul calm or mad, depending on what just happens to be its mood, and if the latter is the case, then I also think of the lovely hours and the little red footstool and whatever else goes together with it.

The days and nights of Whitsuntide I am spending in musical pleasures that God is sending to me by one of those angels who do not menace sensitive souls with the dangerous two-edged sword. This angel is a lady who is already a grandmother and who has kept her soul young and fresh despite the many blows fate had in store for her, and who is marvelously grand and yet truly feminine in her attitudes. In a word, fortunately I do not have the time to torment myself with sweet thoughts on how would it be now if...and if not...etc. I am sure you know how one usually does such things.

Unfortunately, I have not been able to give Byland some of the roses because by that time he had already left for Lenzburg, the lucky fellow. No doubt he will visit you, I asked him to do it. I do not quite understand why, but this gives me some shadow of compensation. I feel as if a part of me had been there -- I know that we see eye to eye in many things. It's funny - as if those eyes had a mission to fulfil -- and this is the very reason I am not coming myself. Now I really must laugh at myself!

Thousand greetings and kisses! Your

Albert

36. FROM MILEVA MARIĆ

[Heidelberg, after 20 October 1897]

It's now been quite a while since I received your letter, and I would have replied immediately, would have thanked you for the sacrifice involved in writing 4 long pages, would have also given some expression to the joy you provided me through our trip together, but you said that I should write to you some day when I happened to be bored. And I am very obedient (you may ask Miss Bachtold), and I waited and waited for the boredom to set in; but so far my waiting has been in vain and I really don't know how to manage this; I could wait from here to eternity, but then you would be right to take me for a barbarian, and, again, if I write, then my conscience is not clear.

I am now roaming, as you have already heard, under German oaks in the lovely Neckartal, whose charms are now, unfortunately, bashfully wrapped in thick pea soup fog, and no matter how much I may strain my eyes, all I see is a certain something, desolate and grey as infinity.

I do not believe that the structure of the human brain is to be blamed for the fact that man cannot grasp infinity, he certainly could do that if in his young days, when he is learning to perceive, the little man wouldn't be so cruelly confined to the earth, or even to a nest, between 4 walls, but would instead be allowed to walk out a little into the universe. Man is very capable of imagining infinite happiness, and he should be able to grasp the infinity of space, I think that should be much easier. And people are so clever, just to think of all they have already accomplished, I see this also here with the Heidelberg professors.

My father gave me some tobacco to take with me and I was supposed to hand it to you, he wanted so much to whet your appetite for our little country of brigands. I talked with him about you, you absolutely must come with me someday. The marvelous conversations you would have here! But I will take over the role of interpreter. But I cannot send it to you, you would have to pay duty on it, and then you would curse me along with my present.

Is it Mr. Sänger who has become a forester? The poor man will probably vent his love in a most romantic Swiss forest. But it serves him right, why does he need to fall in love nowadays, this is an ancient story all the things people know one could spend a lifetime sitting and listening and they would still have more to tell one, everything they have found out themselves. Oh, it was really neat at the lecture of Prof. Lenard yesterday, he is talking now about the kinetic theory of heat of gases; so, it turned out that the molecules of O move with a velocity of over 400 m per second, then the good prof. calculated and calculated, set up eq[uations], differen., integrated, substituted and it finally turned out that even though these molecules do move with this velocity, they travel a distance of only 1/100 of a hairbreadth.

[63]

37. H. F. WEBER'S LECTURES ON PHYSICS

[ca. December 1897 - ca. June 1898]

WEBER'S LECTURES ON PHYSICS

For a sphere of homogeneous material heated to temperature T, whose surface has constantly been kept at the temperature $t = 0$ since time $z = 0$, the law of temperature distribution has been found to be

$$t = t_0 \cdot \frac{2R}{\pi} \frac{\sin \frac{1\pi r}{R}}{r} e^{-\frac{\pi^2}{R^2} \cdot \frac{\kappa}{\rho c} z}$$

$$- t_0 \frac{2R}{2\pi} \frac{\sin \frac{2\pi r}{R}}{r} e^{-\frac{4\pi^2}{R^2} \cdot \frac{\kappa}{\rho c} z}$$

$$+ t_0 \frac{2R}{3\pi} \frac{\sin \frac{3\pi r}{R}}{r} e^{-\frac{9\pi^2}{R^2} \cdot \frac{\kappa}{\rho c} z}$$

This series provides the means for the measurement of κ and at the same time it provides the proof of the previously hypothetically assumed laws of heat flow in an isotropic medium.

If one measures the temperature t at certains times z, then each such observation would provide a means for <measuring> calculating κ, for which the same values would have to be obtained in the different measurements.

r is chosen to be constant in the measurements.

We choose $r = \frac{R}{2}$.

Because of the vanishing of the sine, the 2d, 4th, 6th ... terms of the series drop out. The other sines become alternately +1 and -1. Accordingly, the series reduces to

$$t = \frac{4t_0}{\pi} e^{-\frac{\pi^2}{R^2} \frac{\kappa}{\rho c} \cdot z} - \frac{4t_0}{3\pi} \cdot e^{-\frac{9\pi^2}{<9>R^2} \frac{\kappa}{\rho c} \cdot z} \text{.... ad inf[initum].}$$

One sees that the series converges very quickly because the factors of the terms of the series decrease in the ratio

$$\begin{array}{ccc} 1 & \frac{1}{3} & \frac{1}{5} \\ e^{-1\mu} & e^{-9\mu} & e^{-16\mu} \end{array}$$

For a somewhat large z all subsequent terms are negligible compared with the first one, whence one may set

$$t = \frac{4t_0}{\pi} e^{-\frac{\pi^2}{R^2}\cdot\frac{\kappa}{\rho c}\cdot z}$$

Now we set up the experimental series

$$z_1\text{-------}t_1 = \frac{4t_0}{\pi} e^{-\frac{\pi^2}{R^2}\cdot\frac{\kappa}{\rho c}z_1}$$

$$z_2\text{-------}t_2 = \frac{4t_0}{\pi} e^{-\frac{\pi^2}{R^2}\cdot\frac{\kappa}{\rho c}z_2}$$

$$z_3\text{-------}t_3 = \frac{4t_0}{\pi} e^{-\frac{\pi^2}{R^2}\cdot\frac{\kappa}{\rho c}z_3}$$

$$\ldots \qquad = \qquad \ldots$$

We choose that the time intervals be of equal magnitude and as numerous as possible.

We now have

$$\frac{t_n}{t_{n+1}} = e^{z_n - z_{n+1}} = \text{Constant.}$$

It also can easily be seen that

$$\frac{t_1}{t_m} = \frac{t_2}{t_{m+1}} = \frac{t_3}{t_{m+2}} \ldots = (\text{constant})^m.$$

This conclusion provides us with a means to test the validity of the theory in the most simple way.

For the determination of κ we have

$$\frac{t_1}{t_2} = e^{\frac{\pi^2}{R^2}\frac{\kappa}{\rho c}\Delta z}$$

$$\lg\left(\frac{t_1}{t_0}\right) = \frac{\pi^2}{R^2}\frac{\kappa}{\rho c}\Delta z.$$

The time interval Δz can be chosen arbitrarily from the experimental series. If Δz is taken as the time between 2 readings, then from $n+1$ observations one obtains n equations for the determination of κ.

The definitional units for κ are

minute	cm	1° Celsius	1 gr
Δz	R	c	ρ

For the measurement we use the electrical procedure with the application of a fine solder point. Let A and B be the solder points and G an inserted galvanometer.

gypsum [Fig.]
cooling apparatus

We have then

$$\mathcal{E} = At$$

$$\langle I \rangle i = \frac{\mathcal{E}}{W} = \frac{A}{W}t$$

$$x = \operatorname{tg} x = Bi = \frac{BA}{W}\, t$$

Thus, up to about 10° the angle of deflection can be taken as proportional to the temperature difference.

For objective presentation one can use a concave mirror that is rigidly connected with the mobile magnet. To obtain a line of light one uses a suitable incandescent lamp.

scale [Fig.]

Let the lamp be in A. When the condition $\frac{1}{a} + \frac{1}{b} = \frac{2}{r}$ is satisfied, a real image of the incandescent filament is produced on the scale. If n is the number of scale divisions of length λ, we have

$$\operatorname{tg} 2x = 2x = \frac{n\ \lambda}{D}.$$

Since only quotients of temperatures appear in the expression for κ, the method is a relative one.

$$\frac{t_1}{t_2} = \frac{x_1}{x_2} = \frac{n_1}{n_2}.$$

Thus we have

$$\kappa = \frac{1}{\Delta z}\cdot \lg \left(\frac{x_1}{x_2}\right)\cdot \frac{R^2}{\pi^2}\cdot \rho \cdot c$$

The performed test series yielded the following values for 15-second time intervals:

$$\frac{t_1}{t_2}$$

16.5
15.9
14.5
13
11.7
10.5
9.6
8.7
7.1

The quotients vary between 1.89 and 1.5.
The validity of the law has thus been proved.

Since some materials that are to be tested are difficult to shape into a sphere, the investigation has also to be carried out for the cube, which can easily be produced.

For the cooling of a point of a cube from a distance of $\frac{\alpha}{4}$, where α is the length of the edge, we have

$$t = C'e^{-\frac{3}{4}\frac{\pi^2}{\alpha^2}\frac{\kappa}{\rho c}z}.$$

Thus we have again the constancy of the ratio of two temperatures separated by time Δz.

κ is determined as before.

$$\frac{t}{t'} = e^{+\frac{3}{4}\frac{\pi^2}{\alpha^2}\frac{\kappa}{\rho c}.\Delta z}$$

$$\lg\left(\frac{t}{t'}\right) = \frac{3}{4}\frac{\pi^2}{a^2}\frac{\kappa}{\rho c}\,\Delta z.$$

In this way one finds the coefficients of thermal conduction of rock species.

Granite	*a*	0.525
	b	0.474
Gneiss	*a*	0.577
	b	0.483
Limestone	*a*	0.451
	b	0.404
Sandstone	*a*	0.427
	b	0.181
Glass	*a*	0.125
	b	0.108

a and *b* are randomly picked, non-extreme samples. The conductivity of silicates depends to a great extent on the quartz contained in them, which is a very good conductor. In the case of sandstone, the clay content plays a great role.

In general, the thermal conductivity of substances depends only on the material properties of the components and not on the physical properties of the individual [samples].

The earth in its totality can be considered as a silicate sphere which radiates heat into the low-temperature universe. The insulating effect of the atmosphere can be neglected as small. Hence, assuming that we apply the case of our sphere to the earth, i.e., if we also neglect the heat produced (by contraction and chemical processes), we can state the minimum time necessary to cool the earth by one percent.

$$\frac{t_1}{t_2} = \frac{1}{1-\frac{1}{100}} = 1 + \frac{1}{100} = e^{-\frac{\kappa}{\rho c}\cdot\frac{\pi^2}{r^2}(z_2-z_1)}$$

$$\frac{1}{100} = \frac{\kappa}{\rho\cdot c}\,\frac{\pi^2}{R^2}\cdot \Delta z.$$

$$\kappa = 0.4.$$

The specific heat of the volume = 0.5.
R is the radius of the earth in cm $\pi = 3.1415$.
The calculation yields
1000 million years.

The depth gradient [inverse of the "geothermal gradient"] is usually considered to be a constant number which is independent of the nature of the rock. (By depth gradient we understand the vertical distance between two layers whose temperature difference is 1° Celsius.) This distance is usually taken to be 30 m. This cannot be correct. The depth gradient depends on the internal conductivity of the rock.

surface [Fig.]
rock boundary

We must have

$W_1 = W_2$, since experience shows that at a given depth the temperature is stationary, i.e., heat accumulation does not take place. Hence:

$$\Delta z\,\frac{t_1 - t_2}{h_1}\cdot\kappa_1\cdot f = \Delta z\cdot\frac{t_3 - t_4}{h_2}\cdot\kappa_2\cdot f$$

or

$$\frac{t_1 - t_2}{\frac{h_1}{\kappa_1}} = \frac{t_3 - t_4}{\frac{h_2}{\kappa_2}}.$$

If one assumes that $t_1 - t_2 = t_3 - t_4 = 1°$, then $\frac{h_1}{h_2} = \frac{\kappa_1}{\kappa_2}$. The depth gradients are proportional to the coefficients of thermal conductivity of the rocks.

Diurnal and annual fluctuations of temperature near the surface of the earth.

If one monitors the temperature at the surface of the earth on a normal day throughout 24 hours, one can see that the variation of the temperature can be approximately represented by the formula

$$t = a_0 + a_1 \sin\left[\frac{2\pi}{Z}\cdot z\right].$$

a_0 is the mean daily temperature, $2a_1$ the maximal fluctuation Z = 24 hours.

We wish to find the temperature function for the interior at small depth, while disregarding other changes because of their smallness.

$$dW_1 = -\frac{\partial t}{\partial x}\cdot f\cdot \kappa \cdot dz$$

$$dW_2 = \left(-\frac{\partial t}{\partial x} - \frac{\partial^2 t}{\partial x^2}\,dx\right) f\cdot \kappa\cdot dz \qquad \text{[Fig.]}$$

$$dW_1 - dW_2 = \frac{\partial^2 t}{\partial x^2}\cdot f\cdot \kappa\cdot dx\cdot dz$$

$$dW_1 - dW_2 = \frac{\partial t}{\partial z}\cdot f\cdot \rho\cdot c\cdot dx\cdot dz$$

Hence:

$$\frac{\kappa}{\rho c}\cdot\frac{\partial^2 t}{\partial x^2} = \frac{\partial t}{\partial z}.$$

If we put $u = t - a_0$, then we also have

$$\frac{\kappa}{\rho c}\cdot\frac{\partial^2 u}{\partial x^2} = \frac{\partial u}{\partial z}.$$

For $x = 0$, we must have $u = a_1 \sin \frac{2\pi}{Z}\cdot z$.

$$u = ae^{-mx}\sin\left(\frac{2\pi}{Z}\cdot z - mx\right)$$

$$\frac{\partial u}{\partial x} = -ame^{-mx}\sin\left(\frac{2\pi}{Z}\cdot z - mx\right) - ame^{-mx}\cos\left(\frac{2\pi}{Z}\cdot z - mx\right)$$

$$\frac{\partial^2 u}{\partial x^2} = -am^2e^{-mx}\sin\left(\frac{2\pi}{Z}\cdot z - mx\right) + 2am^2e^{-mx}\cos\left(\frac{2\pi}{Z}\cdot z - mx\right) - am^2e^{-mx}$$

$$\frac{\partial u}{\partial z} = ae^{-mx}\cdot\frac{2\pi}{Z}\cos\left(\frac{2\pi}{Z}\cdot z - mx\right)$$

We must have

$$\frac{\kappa}{\rho c}m^2 = \frac{\pi}{Z} \qquad m = \sqrt{\frac{\pi\rho c}{\kappa Z}}.$$

Hence we have

$$t - a_0 = ae^{-\sqrt{\frac{\pi\rho c}{\kappa Z}}x}\sin\left[\frac{2\pi z}{Z} - <mx> \sqrt{\frac{\pi\rho c}{\kappa Z}}\,x\right].$$

The daily fluctuation at a point x is

$$2\cdot a\cdot e^{-\sqrt{\frac{\pi\rho c}{\kappa Z}}}$$

$$Z = 24\cdot 60$$

$$\kappa\cdot\rho\cdot c = 2.5\cdot 0.2.$$

A fluctuation smaller than $\frac{1}{1000}$ of the fluctuation on the surface we will now consider a negligible quantity.

$$e^{-\sqrt{\frac{\pi\rho c}{\kappa Z}}x} = \frac{1}{1000}$$

$$x\sqrt{\frac{\pi\rho c}{\kappa Z}} = <5>7$$

$$x = ca\ 1.4\text{m}.$$

The agreement with experiment is satisfactory.

Exactly the same consideration holds for annual fluctuations except that for Z the number $365\cdot 60\cdot 24$ is to be introduced here. Here a_0 denotes the mean annual temperature, $2a$ the mean annual fluctuation of the temperature.

Conductivity of metals.

To know the conductivity of metals is of great practical importance because the utility of the metal for a specific purpose often depends on it. However, this conductivity is very variable and can be significantly changed by the presence of even relatively small amounts of impurity.

Krupp examined his metals alloys in the following way:

[Fig.] cooling water

The equation for the temperature on the upper surface is then

$$t = Ce^{-\frac{\pi^2}{h^2}\frac{\kappa}{\rho c}z}.$$

[Note in margin:] One neglects here the heat influx from above.

From that, κ analogous [...] the value of κ thermoelectrically.

Conductivity of liquids and gases: Thermal conduction in liquids.

The measurement of thermal conductivity in liquids is beset by difficulties because heat spreads by motion and mixing of the substance. Weber invented a procedure that makes it easy to determine the constant of thermal conductivity of a liquid or a gas.

copper hole air
liquid [Fig.]
copper
0° cooling water

The copper plates have excellent conductivity (66). They must be produced completely pure (electrolytically). Their temperature can be viewed as being independent of position. [Note in margin: Thermal radiation.] We imagine that the cooling water is acting at time $z = 0$ and neglect the time required by the lower plate to attain the temperature $0^{<o>}$. (The temperature of the cooling water shall be chosen as the 0-point of the scale.) If the <upper plate> liquid does not conduct <temperature> heat, then the upper plate must remain at its temperature because the amount of heat crossing the small glass pieces on which the upper plate is lying can be neglected. However, if the liquid does let the heat pass through, the upper plate is exposed to a temperature which is a function of time.

$$dW = f \cdot \frac{t}{a} \kappa \cdot dz$$ [Fig.]

$$dW' = hFt \; dz$$

$$dW + dW' = \left[\frac{f\kappa}{a} + hF\right] t \; dz$$

$$dW + dW_1 = M \cdot c \cdot -\frac{\partial t}{\partial z} \cdot dz$$

$$\left[\frac{\frac{f\kappa}{a} + hF}{Mc}\right] t <dz> = \frac{\partial t}{\partial z}$$

$$\left[\frac{\frac{f\kappa}{a} + hF}{Mc}\right] z + C = \lg t$$ For $z = 0$

$$C = \lg t_0$$

$$\lg \frac{t}{t_0} = \frac{\frac{f\kappa}{a} + hF}{Mc} z.$$

If one measures a series of t's at Δz intervals, then

$$\lg \frac{\alpha}{\alpha_0} = \frac{\frac{f\kappa}{a} + hF}{Mc} \Delta z = \text{constant}.$$

From this κ.

As it can be seen, if κ is small, then a too must be chosen small to obtain an accurate measurement. It will therefore also be necessary to measure a precisely and to have precisely planeparallel plates.

The former is done using the spherometer. The constant $h \cdot F$ has to be determined by a separate experiment, analogous to the case of the water calorimeter.

Table for κ

Water	0.0816
Glycerol	0.0401
Alcohol	0.0241
Turpentine	0.0150

The value of κ depends strongly on the chemical constitution in the case of liquids as well. Bodies with a complicated molecule have small heat conductivities.

The molecular structure and ρ suffice for predicting the κ of a liquid.

Conductivity of gases.

For the measurement of the conductivity of gases one uses the same apparatus, except that one takes a considerably smaller distance between the two copper plates (about $\frac{1}{10}$ mm). Otherwise the method remains completely analogous. It was found:

Air 0.00360	
Carbon Dioxide	0.00231
Hydrogen	0.0242

What is striking is the <small> large thermal conductivity of hydrogen relative to the other simple gases. Thus hydrogen proves to be akin to metals also in this physical property which depends on the chemical properties.

Thermal insulators.

Though, as a poor heat conductor, air would be well suited to make an excellent heat-insulating medium, it cannot readily be utilized for that purpose because it conveys heat through convection. Hence one uses porous mixtures, finely dispersed solid substances filled for the most part by air, as thermal insulators. Here belong wool, cotton, mineral wool. The latter is suitable for temperatures above 100°, because the silicates, of which it consists, can withstand very high temperatures.

Thermal conductivity depends on the chemical structure of the substance and especially on the fineness of dispersion, which is the reason why the thermal conductivity of the material in question must be separately investigated for each use. This investigation is based on the following relation between the volume of the total mass V and that of the solid substance V_1

$$\frac{V}{\kappa} = \frac{V_1}{\kappa_1} + \frac{V_2}{\kappa_2}.$$

[Fig.]

I.e., no heat accumulation takes place in either of the two parts. If one now assumes that the thermal conductivity is the same as it would be if the two materials would be separated (i.e., that the passage from one substance to the other does not play a significant role), one has

$$W = W_1 = W_2 \quad \text{or} \quad \frac{(t_2 - t_1)\kappa_1}{V_1} = \frac{(t_3 - t_2)\kappa_2}{V_2} = \frac{(t_3 - t_1)\kappa}{V}.$$

$$\frac{(t_3 - t_1)\kappa}{V}\cdot\frac{V_1}{\kappa_1} + \frac{(t_3 - t_1)\kappa}{V}\cdot\frac{V_2}{\kappa_2} = t_3 - t_1$$

$$\frac{V_1}{\kappa_1} + \frac{V_2}{\kappa_2} = \frac{V}{\kappa} \qquad V_2 = V - V_1,$$

Since κ_1 can be found by testing the original material, and κ_2 is known and V is also known, one can determine κ if one knows V_1 or the amount of solid substance contained in a given volume of the mixture. This one determines in the following manner:

[Fig.]

Let the vessel first contain air, which one compresses by appropriately choosing the position of the tube on the right such that the level of mercury is exactly at A. To this corresponds

$$V = \ldots \text{ the pressure } (H+h_1)\rho g.$$

We then reduce the volume to $\frac{V}{m}$ the pressure is then given by $(H+h_2)\rho g$.

By division and application of Mariotte's law one obtains

$$\text{(I)} \qquad \frac{V}{\frac{V}{m}} = \frac{H + h_2}{H + h_1}.$$

Now one brings a certain quantity of the substance to be tested into the space and obtains again 2 analogous observations.

$$V - v \ldots\ldots\ldots\ldots\ldots (H + h_1')\rho g$$

$$\frac{V}{m} - v \ldots\ldots\ldots\ldots\ldots (H + h_2')\rho g.$$

By application of Mariotte's law:

$$\text{(II)} \qquad \frac{V - v}{\frac{V}{m} - v} = \frac{(H + h_2')}{H + h_2'}.$$

By combining I and II one obtains:

$$\frac{1 - \frac{v}{V}}{1 - \frac{mv}{V}} = \frac{H + h_2'}{H + h_1'} \cdot \frac{H + h_1}{H + h_2} = 1 + (m - 1)\,\frac{v}{V}.$$

Thus one obtains $\frac{v}{V}$ and, since V is known, one obtains v. One also has to bear in mind that the correction of Mariotte's law for atmospheric air must not be neglected.

Thus, for example, for $m = 2$, i.e., around 2 atmospheres, the deviation is $\frac{1}{500}$.

$$V_1 P_1 = V_2 P_2 \left(1 - \frac{1}{500}\right).$$

The process of heating and heat loss in pipes.

Rigorous treatment of heat flow from a heated body <occurs> (by way of conduction, convection, and radiation) would be the most difficult problem imaginable. Hence we only set ourselves the task of finding a method that in cases of practical importance permits a quantitative prediction of the heat loss within a few percent. In the case of the water calorimeter we used for the law of heat radiation

$$W = Oh(t - t_a)\,dz$$

and had

$$W = Mc \cdot - dt.$$

Equating these, we get

$$Oh(t - t_a)dz = Mc \cdot - dt$$

$$C + \frac{Oh}{Mc} \cdot z = -\lg(t - t_a)$$

For $z = 0$

$$C = -\lg(t - t_a)$$

$$\frac{Oh}{Mc}\, z = \lg\left\{\frac{t_0 - t_a}{t - t_a}\right\}.$$

This hypothetical law has been confirmed up to a difference of about 5 degrees. For temperature differences of up to 130° we have to add a quadratic correction term. We determine hypothetically

$$dW = h \cdot O\,[(t - t_a) + a(t - t_a)^2]dz.$$

We test the law by observing the cooling of bodies, which we compare with the theory.

In analogy to the above, we have for a body which is cooling off:

$$dW = hO\,[(t - t_a) + a(t - t_a)^2]dz = Mc \cdot - dt.$$

$$C + \frac{hO}{Mc} z = - \int \frac{dt}{(t - t_a) + a(t - t_a)^2} = -\lg \left[\frac{t - t_a}{1 + a(t - t_a)}\right]$$

$$C \qquad\qquad = -\lg \left[\frac{t - t_a}{1 + a(t - t_a)}\right]$$

$$\frac{hO}{Mc} z = \lg \left[\frac{t_0 - t_a}{t - t_a} \cdot \frac{1 + a(t - t_a)}{1 + a(t_0 - t_a)}\right].$$

Thus one has a series of direct measurements for the determination of $h \cdot O$ and a. The values so obtained for the constants must really be constant if the assumption was correct.

Experience shows that a remains constant for different positions, different bodies, different shapes. However $\frac{hO}{Mc}$ varies with all these factors, so that it has to be separately determined for each particular case. $a = 0.0070$.

Under otherwise same conditions h is approximately the same for all polished metals, i.e., is dependent on physical characteristics only. h is large for bodies of great absorptivity (coal).

Pipelines.

3 atm.	130-140°	□ 1 sqcm

$$t - t_a = \text{ca } 130°$$

$$W_{f=1 z=1} = h \cdot 1\, [130 + 0.0070 \cdot 130^2] \cdot 1$$

For iron pipes $h = 0.012$

W = ca 3 cal per <sq>cm^2.

Loss very significant.
One uses a thermoinsulating layer.

We now investigate the reduction of heat losses when the pipe is surrounded by an insulating layer.

[Fig.]

Let the state be stationary. Consequently:

$$W_1 = W = W_2 \quad (z = 1)$$

$$W = -\frac{dt}{dr} \kappa 2r\pi\ell = C$$

$$-dt\kappa 2\cancel{r}\pi\ell = C\frac{dr}{r}$$

$$-t\kappa 2\pi\ell = C \lg r + C'.$$

Holds also for r_1 and r_2.

$$-t_1\kappa 2\pi\ell = C \lg r_1 + C'$$
$$-t_2\kappa 2\pi\ell = C \lg r_2 + C'$$
$$C \lg \left(\frac{r_2}{r_1}\right) = (t_1 - t_2)\cdot 2\,\kappa\pi\ell$$

$$C \lg \left(\frac{r_2}{r_1}\right) = [(t_1 - t_a) - (t_2 - t_a)]\cdot 2\,\kappa\pi\ell.$$

$$C = \frac{(t_1 - t_a) - (t_2 - t_a)}{\lg\left(\frac{r_2}{r_1}\right)}\, 2\kappa\pi\ell.$$

But we also have

$$C = 2h\pi\ell r_2\,[(t_2 - t_a) + a(t_2 - t_a)^2].$$

t_1 is known, hence $t_1 - t_a$ as well, hence $t_2 - t_a$ can be derived from the equality of the two Cs. a is 0.0070 = const.

An experiment was done with an iron pipe that without insulation and at a constant temperature of 130° has a heat loss of 3 cal per sq.cm. The insulation consisted of mineral wool.

The data were as follows:

$$\left.\begin{aligned} \kappa &= 0.049 \\ r_1 &= 5 \text{ cm} \\ r_2 &= 7 \text{ cm} \\ h &= 0.0102 \\ t_1 - t_a &= 130^\circ \\ t_2 - t_a &= x \end{aligned}\right\}$$

$$\frac{130 - x}{\cancel{\lg\left[\frac{7}{5}\right]}}\, 2\kappa\pi\cancel{\ell} = \cancel{2}h\cancel{\pi}\cancel{\ell}7[x - ax^2] \lg\frac{7}{5}$$

$$130 = \left(1 + 7\lg\frac{7}{5}\right)x - 7a\lg\frac{7}{5}\,x^2.$$

$$x = 94 \qquad t_2 - t_a = 94.$$

This yields the heat $C = \dfrac{130 - 94}{\lg\frac{7}{5}}\cdot 2\cdot 0.049\cdot 3.1415\cdot\ell.$

The heat per sq. cm. $= \dfrac{C}{2\pi r_2\ell} = 1.5$ Kcal.

4th. Section.
On the nature of heat.

Thus far "heat" was only the designation of some quantity of an unknown thing regardless of the true nature. This would completely suffice for the explanation of the phenomena if heat were not interrelated with other physical quantities. Now we come to the explanation of these relations.

It has been shown earlier that the work one has to perform to expand a quantity of gas from the volume V_0 to the volume V equals

$$A = - \int_{V_0}^{V} p \, dV .$$

This integral can be evaluated if one knows p as a function of V.

Since the beginning of our century it has been known that the temperature of a gas drops during sudden expansion. Under the tacit assumption that the total amount of heat is invariable, one explained this phenomenon by the assumption that the specific heat is a quantity that varies with the density of the gas.

mV_1 $\qquad$ mV_2

(t_1) $\qquad$ (t_2)

$$m(273 + t_1)c_{V_1} = m(273 + t_2)c_{V_2} = W$$

....

$$\langle V_1\rho_1(273 + t_1)c_1 = V_2\rho_2(273 + t_2)c_2\rangle .$$

Thus it was assumed that the specific heat of a gas at constant volume is a function of density.

The first to call into question the constancy of heat and to perform experiments concerning this matter was Joule.

He used the arrangement shown below:

compressed gas [Fig.]
water

He compressed a certain amount of gas <& that> which was enclosed in the airtightly closed flask by means of 2 stopcocks A and B. If one opened [the stopcock] at B, the gas streamed out through a long capillary. The temperature of the escaping gas was then at all times = the temperature of the water. Let the latter initially be t_0. During the experiment it drops to t.

The outflow of the gas occurs against atmospheric pressure. For the volume V to be able to flow out (at the prevailing pressure P), it has to perform work. This equals

$$P \cdot f \cdot \frac{\Delta V}{f} = P \cdot \Delta V .$$

$$P = H\rho g \qquad \text{[Fig.]}$$

$$A = H\rho g \cdot \Delta V .$$

The temperature in the interior of the vessel is <kept constant during the exp> equalized by a stirrer so that it is a function of time only.

The question now arises: Did the total heat content of the system remain the same during the experiment or did it change during the performance of work A? The thermometer provides a clarification of that, which is placed in the vessel and can be read to $\frac{1}{360}^{\circ}$.

Let the initial temperature in the vessel t_0 agree as closely as possible with the temperature t_a of the surrounding space. If the temperature in the calorimeter decreases to t, the heat removed from the solid and liquid and gaseous masses is

$$(\Sigma(mc)) \cdot (t_0 - t) = W .$$

E.g., $A = 0.703 \cdot 13596 \cdot g \cdot 0.446 = 464\, g\ (\frac{\text{m}}{\text{kg}})$.

[Note in margin:] It is here anticipatively assumed that at constant volume the specific heat of gases is constant regardless of the size of this volume.

This yielded a temperature change of

$$0.1002^{\circ}$$

$$\Delta t' = 0.1002 .$$

To this has to be added one more correction because of the heat released by the vessel to the surroundings.

$W' = h \cdot O \cdot \left[\frac{t_0 + t}{2} - t_a\right] \cdot z$ $\quad hO$ is to be det[ermined] by a separate experiment.

Effect on the temperature change $+ \frac{W'}{\Sigma(mc)}$. In our case the value was

$$-0.010 = \frac{W'}{\Sigma(mc)} .$$

[in margin:] $A = \frac{L}{[g]^2}$

$$\Delta t = 0.0902$$

$W =$ <u>1.060 kg cal.</u>

[in margin:] W

Different experiments showed the proportionality of W and A.

$$\alpha \cdot W = A$$

$$\alpha = 432g .$$

[Note in margin:] Must not depend on g. g is the acceleration of terrestrial gravity that prevailed at that time and that point of the earth!

For all gases α was found to be constant. To a fixed amount of heat lost there always corresponded a fixed amount of work gained.

As mentioned, the old theory explained the phenomenon by assuming that the temperature falls with the expansion of the heat substance, which would expand with the gases. This would produce the work performed, one thought. As long as the specific heat of the gases at constant volume and different densities was not known, it was possible to interpret the above experiment in such a way because it was not possible to ascertain the amount of heat contained in the system.

[Note in margin:] When Joule let his gas flow into an empty space so that it did not have to overcome any pressure, there was no perceptible temperature drop. Irrefutable proof.

Hirn carried out large-scale experiments with steam, proving that heat is indeed being consumed in the cylinder of a steam engine, i.e., supplied and not taken out again. He also determined the equivalent.

[Fig.]

A Steam boiler
B Apparatus for superheating
C Cylinder of the steam engine
D Condenser

We imagine that the engine operates at a steady pace. In unit time, then, the amount of water M leaves the boiler. We follow its caloric fate.

1) Heat intake in A
   $M(t_2 - t_1)$ calories
2) Vaporization in A
   <m> MD (D denotes the heat of evaporation at temp. t_2)
3) Superheating in B
   <m> Mc (c is the specific heat of the saturated vapor)

Then M returns in some state to D. If we supply cooling water of temperature. t_0 such that t_1 stays constant, and if we have to supply μ cooling water per second, then the amount of heat still contained in m will be

$$\mu(t_1 - t_0).$$

Hence the heat lost by m in the cylinder is

$$M(t_2 - t_1 + D + c(t_3 - t_2)) - \mu(t_1 - t_0).$$

On the other hand, it is easy to arrive at the mechanical work performed, e.g., if one knows the dependence of pressure on the momentary position of the piston

[Fig.]

$$\int_{<z>x=0}^{z=1} p\,dx.$$

Hirn found:

$$\alpha = 413.$$

Heat losses which inevitably must arise in the setup through conduction, lead to a smaller-than-calculated heat reserve in front of the piston and to a larger one behind it, so that the actual amount of heat in question is smaller. Therefore the value of α must be somewhat larger than that obtained from Hirn's experiments, which is also evident from Joule's experiment.

Joule was also the first to attempt the quantitative determination of the heat produced during the performance of mechanical work. Joule set up the experimental arrangement shown in the accompanying sketch.

[Fig.]

The weights, which are lifted in the indicated manner, perform upon lowering a completely determined amount of work, which is converted into heat of friction partly in the vessel serving as calorimeter and partly in the axles for the strings. If the experiment is repeated several times, the very sensitive thermometer will show a rise in temperature which permits a conclusion concerning the amount of heat produced. First, one has:

$$2Mgh = A + A' + 2M\frac{v^2}{2}.$$

A denotes the work done in the calorimeter, A' the work lost through friction during transmission, $2M\frac{v^2}{2}$ the kinetic energy present in the weights at the end of the lowering period.

v can be measured directly and the friction in the paddle work is to be chosen so large that v (and A') become small relative to A.

A' is determined in a separate experiment. One places the string around the middle roller in such a way that the two M are in equilibrium. After unscrewing the paddle, one adds to one of the M's a small extra weight m, whose magnitude is such that the system will move with the same speed as in the main experiment. Then one obtains the work $A' = mgh$ required for overcoming the passive resistances in the strings and the transmission axles.

It remains then to demonstrate that the motion brought about in such a system can really be considered as uniform.

It is a law of nature that, within very wide limits, the friction of liquids is proportional to the speed of motion, so that the equation of motion of a weight is

$$M\cdot\frac{d^2x}{dt^2} = Mg - ac.$$

a denotes a proportionality factor which may be increased at will.

$$M \frac{dc}{dt} = Mg - \frac{ac}{M}$$

$$\frac{dc}{g - \frac{ac}{M}} = dt.$$

$$-\frac{m}{a} \cdot \lg \left\{ g - \frac{ac}{m} \right\} = t + C$$

$$-\frac{m}{a} \lg g = C$$

$$-\frac{m}{a} \lg \left\{ 1 - \frac{ac}{mg} \right\} = t$$

$$1 - \frac{ac}{mg} = +e^{-\frac{a}{m}t}$$

$$c = \frac{mg}{a} \left\{ 1 - e^{-\frac{a}{m}t} \right\}.$$

One can see that the larger the a, the faster c approaches the limiting value <m> $\frac{Mg}{a}$.

Thus, the work performed in the calorimeter has now been determined for every variation.

$$A = 2Mgh - A' - 2M \frac{c^2}{2}.$$

We now examine the amount of heat produced.

$$W = \sum (mc) \cdot \Delta t + W'.$$

W' is the quantity of heat lost to the surroundings by conduction and radiation.

$W = Oh \cdot (t - t_a) \cdot z$, where Oh is determined in the familiar manner by auxiliary experiments. If we now define β_0 as a numerical factor such that $A = \beta_0 W$, it was found that one always obtained the same value for β regardless of the substances and weights used. It has been found that $\beta_0 = 425g$. This value was established very carefully from a great number of experimental series.

Joule forgot to introduce a correction that would compensate for the uneven thermal expansion of mercury. We know that $t = t' + \frac{t'^2 - 100t'}{30000}$. Thus, Δt will actually be <larger> smaller, and hence W as well, hence β_0 will be larger. Later on, heat produced by expenditure of a certain amount of <heat> work was determined in another way.

Given an electric conductor of resistance W and current I, the law of heat production derived empirically by Joule reads

$$\mathfrak{W} = \kappa I^2 \cdot W \cdot Z, \text{ where } \kappa \text{ is a constant.}$$

But on the other hand,

$$A = \varepsilon\ I \cdot Z = A = I^2 \cdot W \cdot Z$$

$$A = W \cdot \frac{1}{\kappa}.$$

It has repeatedly been found that

$$\frac{1}{\kappa} = \beta = 427.3 \text{ to } 427.6.$$

The value so found shows satisfactory agreement with that found by Joule.

Analogy of heat and kinetic energy.

First of all, from the assumption that one cannot produce an unlimited amount of work from a cyclic process, it follows that $\alpha = \beta$. I.e., the amount of work arising during the loss of a certain amount of heat is equal to the amount of work that has to be performed in order to regain this same amount of heat.

[Fig.]

$$m \cdot \frac{d^2 s}{dt^2} \langle dt \rangle = K \cdot \cos \nu$$

$$m \cdot \frac{dc}{dt}\ c\ dt = dA \qquad\qquad m \cdot \frac{dc}{dt} = K \cdot \cos \nu$$

$$mc\ dc = dA$$

$$m \cdot \frac{d}{dx} \left[\frac{dx}{dt}\right] \cdot \frac{dx}{dt} = K \cos \nu = \frac{dA}{ds}$$

$$m \left[\frac{c^2}{2}\right]_{c_1}^{c_2} = A$$

$$m \left[\frac{c^2}{2}\right]_{c_1}^{c_2} = \int_{A_1}^{A_2} dA.$$

$$L_2 - L_1 = A.$$

increase on

The decrease in kinetic energy equals the work by [the system].

Having seen that A is also proportional to $\mathfrak{W}$, one can surmise a kinship between L and A or be pushed to the convinction that heat should be viewed as kinetic energy.

[Note in margin]: Potential energy a stepchild.

First of all, the principle of conservation of energy requires that $\alpha = \beta$. We consider the following cyclic process. Using Joule's procedure, we convert a well-determined amount of work A into heat (using the paddle wheel), which we imagine introduced into a closed gas container. The heat is $W = \frac{A_1}{\beta}$. If we now let the gas flow out just until the same amount of heat has been lost again, the heat spent will be $W = \frac{A_2}{\alpha}$. Thus, according to our assumption

$$\frac{A_1}{\beta} = \frac{A_2}{\alpha}.$$

If A_1 and A_2 would differ from each other, it would be possible to obtain A_2 from A_1 without expenditure of heat, and thus gain $(A_2 - A_1)$, which process could be repeated as often as we want. This, however, contradicts the principle of conservation of energy. Because if $A_2 > A_1$, it would be possible to re-use a part of A_2 to reinstate the previous temperature of the gas and to use another part in some other way. The gas would then yield A_2 again, etc. It is necessary that $A_1 = A_2$.

Hence $\alpha = \beta = J = 427g$. This quantity $\alpha = \beta$ shall from now on be called J. J is the work performed during the disappearance of one calorie.

The complete equivalence of the amounts of heat and the amounts of work has now been proved.

The analogy between kinetic energy and heat leads to the mechanical conception of heat.

One starts out from the conception of elastic, freely moving gas molecules, whose kinetic energy shall constitute the heat reserve of the gas.

[Fig.]

According to this assumption, the pressure of the gas arises from the collision forces exerted by the molecules on reflection from the walls of the container. If these walls are in a state of rest, the molecules regain their former velocity. However, if they are receding, the velocity regained by the gas molecules after reflection will be smaller than that before the collision. The work of receding corresponds to the reduction of the sum of kinetic energies in the mass of gas.

Thermodynamics of gases.

Let there be given a gas (mass M). Let its state be characterized by V, t, p. Then there will always hold the general equation of state

$$pV = p_n V_{n_0} (1 + \alpha t).$$

This relation is always valid, regardless of the state of the gas. For an infinitely small change of state it will thus always hold that

$$pdV + Vdp = p_n V_{n_0} \alpha dt.$$

All processes involving a change in the state of gas have to satisfy this equation at all times (if one disregards the correction which depends on the magnitude of the molecular mass v).

However, there is also another equation that can be derived from the equivalence of heat and work:

$$(1) \qquad dW = Mc_V dt + \frac{pdV}{J}.$$

Here c_V denotes the specific heat per unit mass at constant volume of the gas. This quantity is independent of the physical state, as can be surmised from the analogy, as the specific heat of solid substances proved to be a property characteristic of the material only.

[Note in margin:] Can also be inferred directly from Clausius' relation $\sum \frac{mv^2}{2} = \frac{3}{2} pv$!

The 2nd term represents the work performed divided by the equivalent, i.e., also a quantity of heat to be supplied. The expression for dW contains the differentials of two quantities occurring in the general equation of state, and hence 2 additional equivalent equations can be formed for dW by substitution:

$$(2) \qquad dW = Mc_V dt + \frac{p_n V_{n_0} \alpha dt}{J} - \frac{Vdp}{J}$$

$$(2) \qquad dW = \left[Mc_V + \frac{p_n V_{n_0} \alpha}{J} \right] dt - \frac{Vdp}{J} .$$

If p is constant while the amount of heat dW is being supplied, it follows that

$$(2a) \qquad \frac{1}{M} \cdot \frac{dW}{dt} = c_p = c_V + \frac{p_n v_{n_0} \alpha}{J} .$$

One thus obtains the specific heat at constant pressure. v_{n_0} denotes the volume of mass 1.

$$dW = \frac{Mc_V}{p_n V_{n_0} \alpha} (p\ dV + V\ dp) + \frac{pdV}{J}$$

$$(3) \qquad dW = \left[\frac{1}{J} + \frac{Mc_V}{p_n V_{n_0} \alpha} \right] p\ dV + \frac{Mc_V}{p_n V_{n_0} \alpha} V\ dp .$$

If no heat is supplied during a process of change, i.e., $dW = 0$, we have

$$\left[c_V + \frac{p_n V_{n_0} \alpha}{M \langle c_V \rangle J} \right] p\ dV + c_V V\ dp = 0$$

or

$$c_p \cdot pdV + c_V Vdp = 0.$$

This equation represents a relationship between p and V for a process of change during which no heat is being supplied. This is approximately the case with caloric motors [heat engines]. Such a change of state is called *adiabatic*.

Determination of the specific heat of gases.

A calorimetric determination of c_V analogous to the method used for solid substances would be running into great difficulties.

One therefore prefers to determine c_p directly and to derive c_V from that.

We set up the following experimental arrangement:

device for adjusting air stream intensity heating vessel

circular pipe [Fig.]
calorimeter
gas container

The gas stream flowing through the pipe is kept constant with the aid of the adjusting screw and the manometer, such that the air in the heating apparatus takes on a temperature T, which also should remain constant. The calorimeter will then gradually change its initial temperature t_0. In addition, it is assumed that the air leaving the calorimeter has taken on the temperature of the calorimeter. t_0 shall be the temperature of the calorimeter's surroundings.

We want to find the behavior of the temperature of the calorimeter:

$$\sum (mc)\, dt = m_1 \cdot c_p (T - t)\, dz + \kappa \cdot q \cdot \frac{T - t}{a}\, dz - hO(t - t_a)\, dz$$

$$= \left\{ \left[m_1 c_p + \frac{\kappa q}{a} \right] (T - t_a) \langle + \rangle - \left[m_1 c_p + \frac{\kappa q}{a} + hO \right] (t - t_a) \right\} dz$$

$$A \cdot \Sigma (mc) \qquad\qquad B(\Sigma\, mc)$$

$$dt = A - B(t - t_a)\, dz$$

$$-\frac{1}{B} \cdot \lg \left[A - B(t - t_a) \right] = z + C.$$

For $z = 0$ we have $t = t_0$

$$-\frac{1}{B} \lg \left[A - B(t_0 - t_a) \right] = C$$

$$\frac{1}{B} \lg \frac{A - B(t_0 - t_a)}{A - B(t - t_a)} = z\,.$$

Thus the experiment provides for observational data and equations for the determination of A and B. From this A and B can be derived very accurately.

We carry out one more series of tests, such that the velocity of the gas that flows through = 0, but all the other conditions are exactly the same as in the 1st experiment.

[Fig.]

From this we determine the constants A' and B' that correspond to them.

We have now:

$$A = \frac{\left[m_1 c_p + \frac{\kappa q}{a}\right](T - t_a)}{\Sigma(mc)} \qquad B = \frac{\left[m_1 c_p + \frac{\kappa q}{a} + hO\right]}{\Sigma(mc)}.$$

$$A' = \frac{\frac{\kappa q}{a}(T - t_a)}{\Sigma(mc)} \qquad B' = \frac{\frac{\kappa q}{a} + hO}{\Sigma(mc)}$$

$$A - A' = \frac{m_1 c_p}{\Sigma(mc)}(T - t_a) \qquad B - B' = \frac{m_1 c_p}{\Sigma(mc)}.$$

[Note in margin:] In the second series of experiments $m_1 = 0$, all the other constants are the same.

From this, c_p is to be calculated twice, provided that m_1 can be accurately determined. Thus far, all we know about m_1 is that it is constant. m_1 is determined by measurements on the gas container at the beginning and at the end of the test.

Beginning	End
p_1	p_2
t_1	t_1
ρ_1	ρ_2
V	V
M_1	M_2

$$m_1 = \frac{M_1 - M_2}{Z} = \frac{V\rho_1 - V\rho_2}{Z}.$$

However, we have, in general, $\rho = \frac{p}{p_n} \cdot \rho_{n_0} \cdot \frac{1}{1+\alpha t}$.

Hence,

$$m_1 = \frac{V}{Z} \cdot \frac{\rho_{n_0}}{p_n} \cdot \frac{1}{1 + \alpha t}(p_1 - p_2).$$

Now we also know m_1 and hence c_p. If c_p for a given gas is known, then c_V can be calculated from the relation derived above

$$c_p = c_V + \frac{p_n}{\rho_{n_0}} \cdot \frac{\alpha}{J}.$$

c_p depends on the pressure whereas c_V denotes a physical constant that is unconditionally inherent in the body [depending] only on the mass and the chemical characteristic of the substance.

[In the right margin:] ?.

The gases obey a law that is completely analogous to that for the solid bodies:

$$c_V \cdot m = \text{constant}.$$

This holds for O_2, N_2, H_2, CO.
The same holds true for gases whose <atoms> molecules consist of 3 atoms.
Here m denotes the molecular mass.

$$c_V' \cdot m' = c_V'' \cdot m'' .$$

If N denotes the number of molecules per unit mass, we have

$$m' = \frac{1}{N'} \qquad m'' = \frac{1}{N''}$$

$$\frac{c_V'}{N'} = \frac{c_V''}{N''}.$$

The specific heat per molecule has the same value for all equally built molecules.

(Insert)

Using induction, Avogadro found the following law: $\frac{\rho_{n_0}}{m}$ is constant for all gases.

Again, m is the mass equivalent of the molecule.

$$\langle m \rangle \ \frac{\rho_{n_0}'}{m'} = \frac{\rho_{n_0}''}{m''}$$

[Fig.] N molec.

$$\langle m = \frac{1}{N} \rangle$$

$$m'N' = \rho_{n_0}$$

$$\rho_{n_0}' \cdot \frac{N'}{\rho_{n_0}'} = \rho_{n_0}'' \ \frac{N''}{\rho_{n_0}''}$$

$$N' = N''.$$

Under otherwise equal conditions, equal volumes contain equal numbers of molecules.

Adiabatic change of state.

The law derived [above] holds true.

$$c_p p dV + c_V \cdot V dp = 0.$$

For the sake of brevity we denote $\frac{c_p}{c_V}$ by κ, so that κ is a number always > 1.

$$\kappa > 1$$

$$\kappa = \frac{c_V + \frac{p_n v_{n_0} \alpha}{J}}{c_V}.$$

We have then

$$-\kappa p dV = V dp$$

$$-k\frac{dV}{V} = \frac{dp}{p}$$

$$\text{or} \qquad -\kappa \lg V = \lg p + C$$

If the initial state is characterized by $p_1 V_1$,

$$-\kappa \lg V_1 = \lg p_1 + C$$

$$\kappa \lg \left(\frac{V_1}{V}\right) = \lg \left(\frac{p}{p_1}\right)$$

$$\left[\frac{V_1}{V}\right]^{\kappa} = \frac{p}{p_1}. \qquad \text{I.}$$

$$\text{or} \qquad p_1 V_1^{\kappa} = p V^{\kappa}.$$

The analogy of the law for the adiabatic and the isothermal change of state is noteworthy.

The coefficient of elasticity of a gas varies with the kind of the change of state. It has earlier been defined as

$$-V \cdot \frac{dp}{dV} = \frac{\text{increase in pressure}}{\text{decrease in the specific volume}}$$

Isothermal change

$$p = \frac{V_1 p_1}{V} \qquad \epsilon_i = \frac{p_1 V_1}{V} = p$$

Adiabatic change

$$p = \frac{p_1 V_1^{\kappa}}{V^{\kappa}} \qquad \epsilon_a = \kappa \cdot \frac{p_1 V_1^{\kappa}}{V^{\kappa}} = \kappa p .$$

Since κ depends on ρ_{n_0} and c_V, the nature of the gas does play a role here, which was not the case for ϵ_i.

To illustrate this, we present here graphically both changes of state as well as their coefficients of elasticity.

adiabatic change of state [Fig.]
isothermal " " "

To obtain also t for the various values of p and V, we have only to apply the general equation of state

$$pV = p_n V_{n_0}(1 + \alpha t)$$

$$p_1 V_1 = p_n V_{n_0}(1 + \alpha t)$$

$$\frac{pV}{p_1 V_1} = \frac{1 + \alpha t}{1 + \alpha t_1} .$$

Insertion of I into this formula yields a relationship between p and t and between V and t.

$$\left(\frac{V_1}{V}\right)^{\kappa \langle + \rangle - 1} = \frac{1 + \alpha t}{1 + \alpha t_1} \qquad \text{II.}$$

$$\left(\frac{p_{\langle 1 \rangle}}{p_1}\right)^{\frac{\kappa - 1}{\kappa}} = \frac{1 + \alpha t}{1 + \alpha t_1} \qquad \text{III.}$$

If one wants to produce a minimal temperature <in this way> through adiabatic change, this can be achieved by an unlimited increase of V. One obtains

$$\lim 1 + \alpha t = 0 .$$
$$\alpha = -273.$$

Theoretically, the temperature cannot drop further by adiabatic change. (In reality, this limit cannot be reached because of the intervening liquefaction.) In the case of atmospheric air, e.g., this occurs already at -192 degrees.

Direct determination of κ.

Let the following arrangement be set up:

stopcock [Fig.]
gas container

At the start let the pressure in the container be somewhat higher than the atmosph. It has to be measured. $p_1 = (H + h_1)\rho g$. After that one lets the air stream out instantaneously, which means an adiabatic change of state. Let the resulting pressure then be $p_2 = H\rho g$. The stopcock shall be closed immediately after the outflow of the air. The temperature of the air will have been reduced from t_1 to t_2 due to the work performed by it during the outflow of a part of it. Here t_1 shall at the same time be the external temperature t_a. Gradually, however, heat will now flow in from the outside and the pressure will rise to p_3, while the temperature of the gas will become $t_3 = t_1$. $p_3 = (h_3 + H)\rho g$.

According to the foregoing, the following relations hold:

$$\left(\frac{p_1}{p_2}\right)^{\frac{\kappa-1}{\kappa}} = \frac{1 + \alpha t_1}{1 + \alpha t_2} .$$

$$\left(\frac{p_2}{p_3}\right) = \frac{1 + \alpha t_2}{1 + \alpha t_1} \qquad \text{conseq.} \qquad \left(\frac{p_1}{p_2}\right)^{\frac{\kappa-1}{\kappa}} = \frac{p_3}{p_2} .$$

Or

$$\left(\frac{p_1}{p_2}\right)^{\kappa} \cdot \frac{p_2}{p_1} = \frac{p_3^{\kappa}}{p_2^{\kappa}}$$

$$\left(\frac{p_1}{p_3}\right)^{\kappa} = \frac{p_1}{p_2}$$

$$\kappa = \frac{\lg\left(\frac{p_1}{p_2}\right)}{\lg\left(\frac{p_1}{p_3}\right)} = \frac{\lg\left(\frac{H + h_1}{H}\right)}{\lg\left(\frac{H + h_1}{H + h_3}\right)} .$$

We thus have determined $\kappa = \frac{c_p}{c_V}$ by a separate method, because in investigations of adiabatic changes of state this quantity appears in the exponent.

This determination, combined with that of c_p, provides once more a means of testing the value of J.

$$c_p = c_V + \frac{p_n v_{n_0} \alpha}{J}$$

$$\frac{c_p}{c_V} = \kappa \qquad \text{consequently} \qquad c_V = \frac{c_p}{\kappa}$$

$$c_p = \frac{c_p}{\kappa} + \frac{p_n v_{n_0} \alpha}{J}$$

$$J = \frac{p_n \alpha}{c_p \cdot \left[1 - \frac{1}{\kappa}\right] \cdot \rho_{n_0}}.$$

This yields for atmospheric air:

$$J(\text{m kg}) \frac{13596 \cdot 0.760 \cdot 0.00367}{1.293 \cdot 0.2375 \cdot \left[1 - \frac{1}{1.405}\right]} g.$$

One obtains <g> $J = 428g$, a value showing good agreement with the ones found previously.

This is how the first determination of J was done, by Robert Mayer in Heilbronn. He also carried out the following reasoning about the solar heat:

According to experience, a <points> unit area located in the earth orbit receives 2.4 gr.cal. per minute. From this one can calculate the <heat> amount of heat the sun emits per minute. Further, one can easily calculate how long the sun would be able to radiate heat at its present intensity if the solar heat were produced by chemical energy (in terms of our bodies of highest chemical affinity). One finds then that in that case the radiation of heat must have already decreased within historical times, which is not the case, as can be recognized from the plants that were growing in Egypt 5000 years ago, because it is an empirical fact that even a 1% decrease has a great phytogeographic importance. Robert M[ayer] explains this by assuming that the production of heat is brought about by the plunging of small celestial bodies into the sun and the conversion of the kinetic energy of this plunge to heat. He shows that the assumption of a relatively small increase in the mass of the sun would suffice to satisfy its heat requirement.

[Note in margin:] Why doesn't one observe, then, significant fluctuations in the radiation from the sun?)

Processes in our caloric machines

A cyclic process is a process at whose end the substance undergoing it is in the same state as at its beginning.

The physical fate of a fixed mass of gas can be represented by a curve interpreted in the Cartesian system whose coordinates are p and V and to whose points one can affix t, say as an index. If the curve closes upon itself <and is at return>, then we are dealing with a cyclic process. The gas mass studied is completely determined by p_1, V_1, t_1 <when ρ_{n_0} of the gas is known>.

[Fig.]

Let the gas be contracted adiabatically. Its final state will then be characterized by p_2, V_2, t_2. The law derived earlier $p_1 V_1^\kappa = pV^\kappa$, which also defines all intermediary states, will then hold for each instant of time.

The work to be performed in the process is

$$A = -\int_{V_1}^{V_2} p\, dV = -p_1\, V_1^{\kappa} \int_{V_1}^{V_2} \frac{dV}{V^{\kappa}} = +\frac{p_1 V_1^{\kappa}}{\kappa - 1}\left[\frac{1}{V^{\kappa-1}}\right]_{V_1}^{V_2}$$

$$= \frac{p_1 V_1^{\kappa}}{\kappa - 1}\left[\frac{1}{V_2^{\kappa-1}} - \frac{1}{V_1^{\kappa-1}}\right] = \frac{p_1 V_1}{\kappa - 1} \cdot \left[\left[\frac{V_1}{V_2}\right]^{\kappa-1} - 1\right]$$

$$= \frac{p_1 V_1}{\kappa - 1}\left[\frac{1 + \alpha t_2}{1 + \alpha t_1} - 1\right] = \frac{p_1 V_1}{\kappa - 1}\,\frac{\alpha(t_2 - t_1)}{1 + \alpha t_1}.$$

But $\frac{p_1 V_1}{1 + \alpha t} = p_n V_{n_0}$, and therefore

$$A = \frac{p_n V_{n_0} \cdot \alpha}{\kappa - 1}\,(t_2 - t_1).$$

Further:

$$c_p = c_V + \frac{p_n V_{n_0} \alpha}{M \cdot J}$$

$$\kappa = 1 + \frac{p_n V_{n_0} \alpha}{M \cdot c_V \cdot J}; \qquad \frac{p_n V_{n_0} \alpha}{\kappa - 1} = M c_V \cdot J.$$

Hence we have

$$A = M \cdot c_V \cdot J(t_2 - t_1).$$

This is the work which has to be performed during compression. The heat evolved in kilogram calories is thus:

$$W = M c_V (t_2 - t_1)$$

We now consider a cyclic process which can be used for the production of work from heat. We can then set up the following scheme.

	A	W
O_1		W_0
	A_1	
O_2		$W_0 + \frac{A_1}{J}$
	$-A_2$	
O_3		$W_0 + \frac{A_1}{J}$
	0	
O_1		W_0

[Fig.]

If the gas is to be brought first from state O_1 into state O_2 by adiabatic compression, (according to what has been said above) the work to be performed is

$$A_1 = Mc_V \cdot J(t_2 - t_1).$$

The performance of this work is associated with an increase in the heat reserve of the gas, hence

$$W_{(2)} = W_0 + Mc_V(t_2 - t_1).$$

Now let the gas pass to state O_3 by an isothermic change of state. The heat flowing into the gas during the process is spent on overcoming the external pressure.

$$A_2 = \int_{V_2}^{V_1} p\, dV = p_2 V_2 \cdot \int_{V_2}^{V_1} \frac{dV}{V} = p_2 V_2 \lg \frac{V_1}{V_2}.$$

The heat reserve does not increase during this process.

No work is being done during the last transition $O_3 - O_1$. The amount of heat leaving is $Mc_V \cdot (t_2 - t_1)$, so that the heat content [is] again W_0, which is also *a priori* clear.

The sum of the work done by the gas is $-(A_1 - A_2) = A_2 - A_1$. This quantity is represented by the area enclosed by the curvilinear triangle $O_1 O_2 O_3$. The heat that has to be spent on the production of this work equals the heat supplied during the period $O_2 - O_3$, and thus is equal to $\frac{p_1 V_1}{J} \lg \frac{V_1}{V_2} = \frac{A_2}{J}$.

We now define the efficiency of the transfer:

$$g = \frac{\text{work obtained}}{J \cdot \text{heat expended}}$$

$$g = \frac{A_2 - A_1}{J \cdot \frac{A_2}{J}} = 1 - \frac{A_1}{A_2}$$

$$g = 1 - \frac{Mc_V J \cdot (t_2 - t_1)}{p_2 V_2 \lg \frac{V_1}{V_2}} = 1 - \frac{Mc_V J \cdot (T_2 - T_1)}{p_2 V_2 \lg \frac{T_2}{T_1}} (\kappa - 1)$$

$$\left[\frac{V_1}{V_2}\right]^{\kappa - 1} = \frac{1 + \alpha t_2}{1 + \alpha t_1} = \frac{T_2}{T_1} \qquad p_2 V_2 = p_n v_{n_0} \alpha T_2$$

$$g = - \frac{c_V J (T_2 - T_1)\ (\kappa - 1)}{p_n v_{n_0} \alpha T_2 \lg\left[\frac{T_2}{T_1}\right]} + 1$$

$$c_p = c_V + \frac{p_n v_{n_0} \alpha}{J} \qquad c_V(\kappa - 1) = \frac{p_n v_{n_0} \alpha}{J}$$

[109]

$$\frac{(\kappa - 1) J}{p_n v_{no} \alpha} = \frac{1}{c_V}$$

$$g = 1 - \frac{T_2 - T_1}{T_2 \lg \left[\frac{T_2}{T_1}\right]}.$$

The efficiency of the process is thus independent of the nature of the gas, independent of the velocity at which the process is carried out (unless this velocity determined by practical aspects).

The table below provides information on the efficiency of the process at different upper temperature limits

	t_1	t_2	g
(centigrade gas thermometer scale)	0°	50°	8.0%
	0°	100°	14.1%
	0°	150°	19.4%
	0°	200°	23.1%
	0°	250°	26.5%
	0°	300°	29.4%

Since it is hardly possible to choose upper temperature limits higher than 300°, the maximal theoretically possible efficiency is 29.4%.

Derivation of the 2nd law of the mechanical theory of heat.

Carnot's cyclic process.

Carnot set up the following cyclic process:

$O_1 - O_2$ (adiabatic, $O_2 - O_3$ (isothermal)

$O_3 - O_4$ (adiabatic, $O_4 - O_1$ (isothermal).

This cyclic process can be used for the conversion of heat into work and vice versa.

First we study this process for gases and we study the amounts of work and heat arising during it.

	Heat content	Heat supplied	Work done by the gas
O_1	W_0	$W_1 =$	$A_1 = -JMc_V(t_2 - t_1)$
O_2			$A_2 = p_2 V_2 \lg \left[\frac{V_3}{V_2}\right]$
O_3			$A_3 = JMc_V(t_2 - t_1)$
O_4			$A_4 = -p_4 V_4 \lg \left[\frac{V_4}{V_1}\right]$
O_1	W_0		

[Fig.]

$A_1 = \int_{V_1}^{V_2} p dV$. Here $p_1 V_1^\kappa = pV^\kappa$ holds

$$= p_1 V_1^{\kappa} \int_{V_1}^{V_2} \frac{dV}{V^{\kappa}} = - \frac{p_1 V_1^{\kappa}}{\kappa - 1} \left[\frac{1}{V_2^{\kappa-1}} - \frac{1}{V_1^{\kappa-1}} \right] = - \frac{p_1 V_1}{\kappa - 1} \left[\left[\frac{V_1}{V_2}\right]^{\kappa-1} - 1\right]$$

$$= - \frac{p_n V_{n_0} (1 + \alpha t_1)}{\kappa - 1} \left[\frac{T_2}{T_1} - 1\right] = - \frac{p_n \langle V \rangle v_{n_0} \alpha T M}{\kappa - 1} (T_2 - T_1)$$

$$\frac{c_p}{c_V} - 1 = \frac{p_n v_{n_0} \alpha}{J}$$

$$\kappa - 1 = \frac{p_n v_{n_0} \alpha}{J c_V}$$

$$A_1 = -JMc_V (t_2 - t_1).$$

Analogously, we get for A_3

$$A_3 = JMc_V (t_2 - t_1), \text{ hence}$$
$$A_1 = -A_3 .$$

Thus, A_1 and A_3 are equal and opposite. The work during the adiabatic change of state is thus proportional to the temperature difference.

We now calculate A_2 and A_4. These works correspond to isothermal changes.

$$A_2 = \int_{V_2}^{V_3} p dV = \quad , \text{ since } pV = p_2 V_2$$

$$= p_2 V_2 \int_{V_2}^{V_3} \frac{dV}{V} = p_2 V_2 \lg \left[\frac{V_3}{V_2}\right].$$

Analogously it follows that

$$A_4 = -p_4 V_4 \lg \left(\frac{V_4}{V_1}\right).$$

The heat inputs in the 2d and 4th phase are defined by the condition that the changes of state shall be isothermal, or, what amounts to the same, since c_V is strictly a characteristic of the substance and hence the heat content is also determined by the temperature only: the heat content must remain unchanged. Thus, it follows from the law of the equivalence of heat and work that to a performed work dA there must correspond a heat input $\frac{dA}{J}$. Hence the quantities of heat W_2 and W_4 which have to be supplied equal the [corresponding amounts of] work divided by J.

As far as the heat content of the gas at the end of the phases is concerned, it is to be said that

1) during adiabatic change the increase in heat is $= \frac{-A}{J}$

2) during isothermal change " " " " $" = 0$.

Now we are able to construct the table.

	W contained	W supplied	Work performed
O_1	W_0		
		0	$-JMc_V(t_2 - t_1) = A_1$
O_2	$W_0 + Mc_V(t_2 - t_1)$		
		$\frac{p_2V_2}{J} \lg\left[\frac{V_3}{V_2}\right]$	$p_2V_2 \lg\left[\frac{V_3}{V_2}\right] = A_2$
O_3	$W_0 + Mc_V(t_2 - t_1)$		
		0	$JMc_V(t_2 - t_1) = A_3$
O_4	$<Mc_V(t_2 - t_3)> W_0$		
		$-\frac{p_4V_4}{J} \lg\left[\frac{V_4}{V_1}\right]$	$-p_4V_4 \lg\left[\frac{V_4}{V_1}\right] = A_4$
O_1	W_0		

Special remarks.

1) Work performed

$$A = (A_2 + A_3) + (A_1 + A_4) = (A_2 + A_4).$$

We rewrite this expression.

$$p_2V_2 \lg \left[\frac{V_3}{V_2}\right] - p_4V_4 \lg \left[\frac{V_4}{V_1}\right]$$

$$\left.\begin{aligned} p_2V_2 &= p_nV_{n_0}\alpha T_2 \\ p_4V_4 &= p_nV_{n_0}\alpha T_1 \end{aligned}\right\} \qquad \frac{V_3}{V_2} = \frac{T_2}{T_1}\frac{V_4}{V_1}.$$

The reason for this rel[ationship] between the 4 volumes is as follows. We have

$$\left[\frac{V_1}{V_2}\right]^{\kappa-1} = \frac{1 + \alpha t_2}{1 + \alpha t_1} \qquad \left[\frac{V_3}{V_4}\right]^{\kappa-1} = \frac{1 + \alpha t_1}{1 + \alpha t_2}$$

$$\frac{V_1}{V_2} = \frac{V_4}{V_3} \qquad \text{or} \qquad \frac{V_4}{V_1} = \frac{V_3}{V_2}$$

Hence

$$A = A_2 + A_4 = p_nV_{n_0}\alpha \lg \left(\frac{V_3}{V_2}\right) (T_2 - T_1).$$

The ratio between the work produced and the work that corresponds to the heat supplied is $\frac{A_2 + A_4}{A_2}$.

or $$\frac{p_n V_{n_0} \alpha \lg \left[\frac{V_3}{V_2}\right] (T_2 - T_1)}{p_n V_{n_0} \alpha \lg \left[\frac{V_3}{V_2}\right] T_2} = 1 - \frac{T_1}{T_2}.$$

Or also - The ratio of the heat converted to work and the heat utilized (the efficiency of work production g)

$$g = 1 - \frac{T_1}{T_2}.$$

Thus, here too the efficiency depends on the temperatures alone.
2) We consider, further, the ratio of the work produced to the work that corresponds to the escaped heat W_4.

$$\frac{A}{JW_4} = \frac{p_n V_{n_0} \alpha \lg \left[\frac{V_3}{V_2}\right] (T_2 - T_1)}{p_n V_{n_0} \alpha \lg \left[\frac{V_3}{V_2}\right] T_1} = \frac{T_2 - T_1}{T_1}.$$

[Note in margin:] Thus, $\frac{A}{W_4} = \frac{J(T_2 - T_1)}{T_1}.$

3) We consider the integral

$$\int \frac{dW}{T} \text{ throughout the whole process.}$$

We divide it into 4 parts according to the phases of the process; in the first part dW is always 0 and so it is in the 3d. It only remains to examine the phases 2 and 4. Since here the temp. is constant, we have

$$\int \frac{dW}{T} = \frac{\int_{O_2}^{O_3} dW}{T_2} + \frac{\int_{O_4}^{O_1} dW}{T_1} = \frac{1}{J} \left[\frac{A_2}{T_2} + \frac{A_4}{T_1} \right]$$

$$= \frac{1}{J} p_n V_{n_0} \alpha \lg \left[\frac{V_3}{V_2}\right] \left[\frac{T_2}{T_2} - \frac{T_1}{T_1}\right] = 0 .$$

The integral $$\int \frac{dW}{T} = 0$$

for the Carnot process. This result is generally valid for every reversible cyclic process.

If the gas is passed through all the states in the opposite direction, the nature of the processes does not change at all, except that always work production takes the place of work consumption, and heat loss takes the place of heat gain. Hence, the above table, read from bottom to top, provides the respective data for this case as well, if one reverses all the signs of heat inputs and work productions.

<4> It can be seen that the following relation holds true:

$$W_2 = W_4 + \frac{A}{J}$$

or. The heat lost during the cyclic process, or the excess of the heat supplied over the heat taken out is equivalent to the work produced.

One can imagine this process to be realized in the following way:

[Fig.]

We imagine that during the isothermal phases O_2O_3 and O_4O_1 the gas is in contact with two well-conducting bodies K_2 and K_4, whose temperatures shall be t_2 and t_1, respectively, during the entire process.

In that way it can be achieved that the processes $O_2 - O_3$ and $O_4 - O_1$ will be indeed isothermal. Regarding K_2 and K_4 it should be noted: If the gas does some work, then it transfers a certain amount of heat from the warmer to the colder body.

However, if we reverse the process, then heat will be removed from the colder body K_4 to compensate for the work $O_1 - O_4$ and delivered to K_1 because the work O_3O_2 is negative; in this case, however, there must be some external work produced, or the work of the gas is negative.

Thus we see: If heat is converted to work by the Carnot process, then heat flows from the warmer to the colder level.

Now we have to apply this consideration to arbitrary bodies.

Within narrow limits, an arbitrary substance obeys the law

$$p - p_1 = \frac{1}{\beta V_1}(V_1 - V). \qquad \text{[Fig.]}$$

β is a coefficient which varies with the nature of the substance. β also depends on whether the volume change is instantaneous, i.e., adiabatic, or slow (isothermal).

A <change> reduction of the volume is accompanied by performance of work, i.e., production of heat, which in turn tends to expand the body, i.e., causes an increase of the pressure. Thus, there exist two different values of β, i.e.,

β_i for isothermal change of state

β_a for adiabatic " " " .

They always obey the relation

$$\beta_i > \beta_a .$$

The curves which represent the change of state in the $(p\text{-}V)$ plane in this case are straight lines. Once again, the adiabatic curve will have the steeper slope against the V-axis.

[Fig.] Adiabatic
Isotherm.

In this case the Carnot process can be carried out farther. The work performed will again be represented by the area of the quadrangle and is again positive when the cycle is traversed in the direction indicated, and negative in the opposite case.

[Fig.]

Once again, heat is absorbed during the path $O_2 - O_3$ and released between O_4 and O_1, while just the opposite is true for the reverse process.
We now imagine that the process extends over infinitely small changes of state and calculate the quantities A and W_4.

$$A = dV \left[\frac{\partial p}{\partial t}\right]_V dt$$

$$W_4 = \left[\frac{\partial W}{\partial V}\right]_t dV$$

[Fig.]

$$\frac{A}{W_4} = \frac{\left[\frac{\partial p}{\partial t}\right]_V dt}{\left[\frac{\partial W}{\partial V}\right]_t}.$$

We now combine two cyclic processes; let the one be carried out by a gas, and the other by some other arbitrary body. Let the work A be produced during the cyclic process I, and consumed during the process II. Further, let the bounding temperatures t_1 and t_2 be the same in the two cases. Further, let both of them refer to infinitely small changes of state.

Let the acting substance be a gas in the cyclic process I, and an arbitrary substance in II.

We now claim that necessarily $W_4 = W_4'$.

[Fig.]

If we would have $W_4' > W_4$, that would mean that heat is being transferred from the colder body K_4 to the warmer body K_2 without the necessity of expending work. $W_4' - W_4$ would be taken from the lower body K_4 and
$W_2' - W_2 = \left[W_4' + \frac{A}{J}\right] - \left[W_4 + \frac{A}{J}\right] = W_4' - W_4$ would be delivered to the upper body. If $W_4' < W_4$, it would only be necessary to reverse both cyclic processes in order to arrive at the result that in this case as well it would be possible to transfer heat from K_4 to K_2 without expenditure of work. However, this is in conflict with experience. The following empirical law holds true:

Without expenditure of work it will never be possible to transfer an amount of heat from a body at a lower temperature to a body at a higher temperature. We have therefore

$$\frac{A}{W_4} = \frac{A}{W_4'}$$

<Or> According to the above, <since> it holds that

$$\frac{A}{W_4} = \frac{J(t_2 - t_1)}{T_1} = \frac{JdT}{T}$$

$$\frac{A}{W_4 T} = \frac{dV\left[\frac{\partial p}{\partial t}\right]_V dt}{dV\left[\frac{\partial W}{\partial V}\right]_t} = \frac{\left[\frac{\partial p}{\partial T}\right]_V}{\left[\frac{\partial W}{\partial W}\right]_t} dT.$$

Hence, quite generally

$$\frac{J}{T} = \frac{\left[\frac{\partial p}{\partial T}\right]_V}{\left[\frac{\partial W}{\partial V}\right]_t}.$$

This is the second law of the mechanical theory of heat.

Saturated vapors.

If a liquid is brought into contact with an empty space, a certain amount of it evaporates into this space. That is to say, the evaporation stops after a certain time, so that after a sufficiently long time period, at a specific temperature equilibrium is established.

[Note in margin:] The vapor in con[tact] with the mother liquid is called saturated.

The next thing to do is to determine the pressure of the saturated vapors in each particular case, depending on the variation of substance, temperature and volume.

Measurement of the pressure.

Let us set up the arrangement shown below. The equation yields

$$p + h_m \cdot \rho g = h\rho_q \cdot g$$
$$p = g[h\rho_q - h_m\rho_m].$$

[Fig.]

mother liquid
mercury

The 2d term in the brackets is usually small compared with the first one, and can be neglected in most cases. If the tube 2) can be shifted vertically, then the volume of the saturated vapor can be changed at will. If this is done slowly enough so that an equalization of temperature between the gas and the surroundings can always be accomplished, and one measures the temperatures associated with the different volumes, one finds that the pressure of a saturated vapor is independent of its volume.

<If one changes> However, if one raises the *temperature*, the

descent of the column shows immediately that the pressure of the saturated vapor has increased. The method of measurement described is adequate for the measurement of <atmospheric> pressures of saturated vapors up to the atmospheric pressure.

At equal <atmospheric> temperatures, different substances show completely different pressures of their saturated vapor. A precise investigation by Regnault has shown that even ice has a vapor pressure. From this one concludes that even solid substances can pass directly into vapor, without having gone through the liquid phase. This is also proved sufficiently by the odor of solid substances.

The mechanical theory of heat explains the fact that vapor pressure is independent of the volume in the following way.

According to Villarceau's rule one has to calculate $A = \sum m \frac{v^2}{2} = \frac{3}{2}pV$. This holds for every movable system that is not subjected to external forces other than the forces of pressure and whose molecular forces are being neglected.

If one sets now $m_1 = m_2 \ldots = m$

$$\& \qquad v_1 = v_2 \ldots = v$$

one obtains

$$Nm \frac{v^2}{2} = \frac{3}{2}pV$$

$$p = \frac{2}{3} \cdot \frac{N}{V} \cdot m\frac{v^2}{2}.$$

If one puts the number of molecules in volume $1 = n_1$,

$$p = \frac{2}{3}n_1 \cdot m\frac{v^2}{2}.$$

That means: If one assumes that equal volumes contain an equal number of molecules, one obtains for p an expression that is independent of the volume V, which of course it is. If one further sets $T = a\,\frac{mv^2}{2}$, where a is an arbitrary constant, then one obtains

$$p = \frac{2}{3}a \cdot n_1 \cdot T.$$

Hence, if n_1 [were] independent of T, then p and T would be proportional. However, it can easily be seen that n_1 increases with T. For n_1 depends first of all on the cohesion of the liquid, which, as we know, is a function of temperature. [In the left margin:] ?

Since it has not yet been possible to calculate the quantity n_1, the only thing one can do is to determine the function between p and T empirically.

Since the method described earlier only permits the measurement of pressures up to one atmosphere, a general method for measuring the pressure of saturated vapors is to be given first.

The process of boiling.

Let us consider a liquid that has absorbed atmospheric air and whose absorptivity decreases with temperature. If we heat the liquid, the excess dissolved air will be removed from everywhere within the liquid in the form of little bubbles.

[Fig.]

For such a little bubble to be stable, the pressure of the gas exerted from the inside on its boundary must be equal to the atmospheric pressure. However, evaporation of the liquid occurs at the surface of the bubble as well, such that the bubble is constantly saturated with the vapor of the liquid (we anticipate the law that the pressures of different substances are simply additive.)

The equilibrium condition applying to the bubble is therefore

$$p_g + p_d = P \qquad P = 760\rho g$$

$$p_g = \frac{p_n V_{n_0}(1 + \alpha t)}{V}$$

$$V_{n_0} = M \cdot \langle\rho\rangle v_{n_0} = \frac{M}{\rho_{n_0}}$$

$$p_g = P - p_d \langle d \rangle = \frac{M}{\rho_{n_0}} \cdot \frac{p_n}{V} \cdot (1 + \alpha t).$$

From this one obtains

$$V = \frac{Mp_n}{\rho_{n_0}} \cdot \frac{1 + \alpha t}{P - p_d}.$$

Boiling sets in when no volume of the bubble satisfies this equilibrium condition any longer; but then we have precisely $p_d = P$.

During boiling the external pressure equals the pressure of the saturated vapor.

The correctness of this explanation is shown by the fact that boiling becomes scanty after it has been going on for a long time, i.e., after most of the gas has escaped, and by the fact that it becomes lively again the moment more gas is introduced in one way or other.

We use the boiling process to determine the relation between t and p_d for an arbitrary liquid whose absortivity for atmospheric air (or some other gas) decreases with increasing temperature.

One sets up the following experimental arrangement.
With this apparatus one measures a long series of temperatures and the corresponding pressures P. The P is then always equal to the vapor pressure p_d. In this way one obtains empirically the function between

t and p_d for a given liquid.

[Fig.] mercury

Since this function [is] very important in the case of steam, it is worthwhile to memorize the following approximate relation:

t	p_d in atm.
100	1
120	2
160	6
180	10

[Fig.]

One sees from this that the temperature varies much more when the pressure varies at low pressures than at high pressures.

Regnault [carried out] a more precise investigation of the function using the following approximation formula fitted to the nature of the curve:

$$\log p = A + B\alpha^{t+20} + C\beta^{t+20}.$$

From a long series of experiments he found:

$$A = 6.26$$
$$B = -1.379$$
$$C = -4.925$$
$$\alpha = 0.98639$$
$$\beta = 0.99619.$$

Differentiation and expansion in a series yields:

$$\not{\frac{1}{p}} \cdot \frac{dp}{dt} = p\,[0.0715 - 0.000462\,\mathrm{t} + 0.00000105\,t^2].$$

Experimental determination of ρ and v.

[Fig.]

Let the inner vessel contain a certain mass m <of a> of the liquid to be studied. The external vessel [shall contain] a significantly larger amount of the same liquid. The two vessels are connected by a mercury manometer as indicated. Care has to be taken that the two vessels be of the same temperature at all times. According to what has been said earlier, this results in the formation of saturated vapors of equal pressure. Consequently, the two little mercury columns of the manometer maintain the same height. One now slowly heats the whole system. The density of the evolved vapor will then increase, while the quantity of liquid in both vessels will decrease. The manometer remains in a state of equilibrium during all this time. This continues until the liquid of smaller mass in the inner vessel has completely been used up. At that point the right column of the manometer rises instantaneously.

The reason for this is as follows: As long as liquid was available in both vessels, the increase in pressure, which corresponded to a certain increase in temperature, was in both of them produced by two factors:

1) by the increase in the temperature of the vapor already present, with the volume remaining the same (if no new quantities of vapor were added).

2) by the decrease in the volume of the old quantity of vapor owing to the newly formed vapors. Obviously, these two factors are additive.

If now the liquid in the inner vessel is exhausted, then a further temperature increase Δt will correspond to a smaller increase of p than in the other vessel, because the 2nd reason for Δp is absent. Thus, the equilibrium of the manometer becomes perturbed.

At this critical temperature we thus have

$v_t = \frac{V}{m}$, where V is the volume of the 2nd vessel. If one takes different masses of m into the small vessel, then one obtains ever more completely empirically the relation between

t and v or between t and ρ

E.g., for water at 100°

<V>v = 1.64 (kg cm^3), cons[equently] ρ = 0.61.

If we now summarize, it follows that for a fixed temperature t and a particular liquid p, v, (ρ) are completely determined.

An *a priori* theoretical determination of v fails because the actual cohesive forces of the liquid molecules that correspond to a given temperature are not known, and according to the mechanical theory of heat v must depend on them.

We now examine what will happen if we have a liquid whose temperature is below the boiling point and we introduce saturated vapor of the same liquid into it.We examine the equilibrium of a little bubble.

[Fig.]

Its walls are acted upon by the atmospheric pressure p from the outside. From the inside out, the vapor pressure of the liquid acts upon the unit surface. Since the temperature of the liquid and, consequently, of the bubble, too, is below the boiling temperature of the liquid, the internal vapor pressure is smaller than the atmospheric pressure. Since we assume that no other forces act upon the system, we must conclude that the bubble cannot be in equilibrium; rather, it must steadily decrease in volume until it disappears altogether. The accompanying pictorial sketch shows the design of the experiment with H_2O.

[Fig.]

Once it has been determined, the relation between t and p allows the measurement of the temperature, which is the more accurate the smaller the temperature change that corresponds to a given change in pressure. This condition is obviously the better satisfied the lower the temperature at which the pressure increases from 0 to P, i.e., the lower the boiling temperature.

E.g., if one chooses SO_2, the boiling temperature is approximately . In the 0 - 20° range, for $\Delta t = \frac{1}{50°}$ one aims at a 1-mm change of a mercury column.

$$p_d = h \cdot \rho_q \cdot g.$$

From this t, using the function between p and t.

[Fig.]

Heat of evaporation of a liquid. D
Formation of saturated vapor.

By D we understand the <vapor qu> quantity of heat necessary to convert the mass 1 to saturated vapor at a given temperature. Since the temperature remains unchanged during the vaporization, one has to supply an amount of heat which is exactly equivalent to the sum of the internal and external work performed.

The external work during the vaporization of the mass is due to the volume increase $v - v_1$, where v is the sp[ecific] v[olume] of the vapor, and v_1 that of the liquid against the external pressure, which equals the vapor pressure p.

Thus, the external work is:

$$p \cdot (v - v_1).$$

The heat that has to be supplied for its production is

$$\frac{p(v - v_1)}{J}.$$

The internal work consists in an increase of the intermolecular distance.

[Note in margin:] Investigate! Vacation.

However, since the function of attraction and of the intermolecular distance is not known, this work A_1 cannot be determined theoretically.

Thus, we now have

$$D = \frac{A_1}{J} + \frac{p(v - v_1)}{J}.$$

For water at 100°, e.g.,

$$v_1 = 0.001$$
$$v = 1.640.$$

Thus, the volume increases in the ratio

$$1 : 1640.$$

The D of a given liquid is a function of temperature. Each time t is varied upwards, A_1 will certainly decrease, because the initial distance between 2 molecules becomes larger and the final distance smaller. p grows very fast while $(v - v_1)$ decreases.

For water, for example, $D = 596 - 0.60\, t$, thus it decreases with t, which shows that the change in A_1 has <more> a great effect.

By considering the reverse process it is easy to demonstrate that the heat of condensation equals the heat of evaporation.

Experimental determination of D.

Let us set up the following experimental arrangement:

[Fig.]

saturated vapor

liquid to be investigated

$T_{constant}$

city gas

The two calorimeters on the left and on the right are exactly alike and are connected with the middle pipe in the same way. As soon as the temperature of the liquid has reached the level for which we want to examine D, we make sure that the external pressure P is the same as the gas pressure of the liquid and we open the stopcock to the calorimeter on the right. The saturated vapor of temperature T will then be brought to the temperature t of the calorimeter, and will be almost completely condensed in the process. We arrange that this condensation be as complete as possible (the higher the T the better), and assume the aim has been achieved.

The formation of vapor shall proceed uniformly.

If m_1 denotes the mass of vapor that flows into the calorimeter on the right in unit time, then the heat corresponding to the time differential dt is

$$D \cdot m_1 \, dz .$$

We now formulate the heat input for the calorimeter on the right

$$dW = m_1 D dz + m_1 c(T - t)dz + L(T - t)dz - H(t - t_a)dz .$$

condensation cooling heat conduction dissipation

On the other hand:

$$dW = \sum (mc) \cdot dt .$$

This gives

$$\sum (mc)\, dt = \Big[\{m_1 D + m_1 c(T - t_a) + L(T - t_a)\} <-H(>$$
$$- \{m_1 c(t - t_a) + L(t - t_a) + H(t - t_a)\} \Big] dz$$

$$\sum (mc)\, dt = \Big[\{m_1 D + (m_1 c + L)(T - t_a)\} - \{(m_1 c + L + H)(t - t_a)\}\Big] dz$$

$$dt = \Big[A - B(t - t_a)\Big] dz$$

$$z = -\frac{1}{B} \lg (A - B(t - t_a) + C.$$

For $z = 0$, we have $t = t_0$

$$C = \frac{1}{B} \lg (A - B(t_0 - t_a)$$

$$z = \frac{1}{B} \lg \{ \frac{A - B(t_0 - t_a)}{A - B(t - t_a)} \}.$$

An analogous result is obtained for the other calorimeter, except that the terms in A' and B' change in such a way that $m_1 = 0$. Now one reads the temperatures for a certain number of time points at equal intervals, and thus obtains a definite relation between z and t for each <thermom> calorimeter.

[Fig.] left

[Fig.] right

From these, the still undetermined constants A and B can be calculated with any desired accuracy, and so can A' and B'. However, we have:

$$A = \frac{m_1 D + m_1 \cdot c(T - t_a) + L(T - t_a)}{\Sigma(mc)}$$

$$A' = \frac{L(T - t_a)}{\Sigma(mc)}$$

$$A - A' = \frac{m_1 D - m_1 \cdot c(T - t_a)}{\Sigma(mc)} .$$

c of the liquid is to be considered known, and so is $\Sigma(mc)$ of the calorimeter. It remains to determine m_1.

$$m_1 \cdot Z = M$$
$$m_1 = \frac{M}{Z}$$

where M denotes the total quantity of the evaporated liquid, which is found by weighing, and Z denotes the duration of the experiment.

For water we have, as mentioned earlier:

$$D = 596 - 0.60\ t. \quad \text{(Approximately)}.$$

Such functions are obtained empirically, by determining D for many T's.

Application of the second law of the mechanical theory of heat to saturated vapors.

We found for an arbitrary substance

$$\frac{J}{T} = \frac{\left[\frac{dp}{dt}\right]_{v}}{\left[\frac{\partial W}{\partial v}\right]_{t\ const.}} .$$

$\frac{dp}{dt}$ has been calculated above for water. $\frac{\partial W}{\partial v}$ *is the quantity of heat necessary to produce a volume increase 1*, hence

$$\frac{\partial W}{\partial v} = \frac{D}{(v - v_1)}.$$

Thus, the law

$$\frac{J}{T} = \frac{\frac{dp}{dt}(v - v_1)}{D} \qquad \text{I.}$$

is universally valid for saturated vapors.
One can see that, according to the above, the quantities on the right-hand side are known as functions of T. Hence we get a means to derive the quantity J for every substance and every temperature, which can also serve as the touchstone of correctness for the functions p, v, and D.

For water at the temp. of 100° one gets, e.g., (with m and kg and seconds as units)

$$T = 373 \quad v_{100} = 1.640 \quad v_1 = 0.001.$$

$\frac{\partial p}{\partial t}$ can be obtained either graphically from the vapor pressure curve or by calculation

t	p		
100	760	27.8	Average 27.85 (mm Hg).
99	733.2		
101	787.9	27.9	

Thus we have

$$J = \frac{373 \cdot 1.639 \cdot 27.85 \cdot 13596\ g}{536}$$

$$J = 427\ g$$

(Not checked). We now rewrite the equation further.

$$\frac{\frac{1}{p} \cdot \frac{dp}{dt} (v - v_1)p}{D} = \frac{J}{T}.$$

Here v_1 is omitted because of its relative smallness.

$$pv = \frac{J}{T} \cdot \frac{D}{\frac{1}{p} \cdot \frac{dp}{dt}}$$

Here $\frac{1}{p}\frac{dp}{dt} = c_1 + c_2 t + c_3 t^2$.

We found for water, e.g.,

$c_1 = 0.0715 \quad c_2 = -0.000462 \quad c_3 = 0.00000103.$

Thus we have the above equation:

$$\frac{p}{\rho} = \frac{J}{T} \cdot \frac{D}{c_1 + c_2 t + c_3 t^2} \qquad \text{II.}$$

This equation is called the general equation of state of saturated vapors because its form is analogous to the general equation of state for gases:

$$\frac{p}{\rho} = \frac{p_n}{\rho_{n_0}} (1 + \alpha t).$$

The left-hand sides are analogous while in both cases there appears a function of temperature alone on the right-hand side. However, in the case of saturated vapors the *nature* of this function depends on the nature of the substance. Besides, in the case of saturated vapors, this equation cannot have the same fundamental importance because p and ρ are individually determined functions of t. A further conclusion of importance is that the law permits a simplification of the expression for the molecular work A_2 performed during evaporation.

$$\frac{J}{T} = \frac{\frac{dp}{dt}(v - v_1)}{D}$$

$$\frac{p(v - v_1)}{J} = \frac{D}{T} \cdot \frac{1}{\frac{1}{p}\frac{dp}{dt}}.$$

But earlier we found

$$D = \frac{p(v - v_1)}{J} + \frac{A_2}{J}.$$

Here p denotes the external pressure. However, in the formula above p denotes the <gas> vapor pressure <of>.

If one imagines that the saturated vapor forms according to this formula, one sees that the pressure of the saturated vapor formed must be equal to the external pressure (i.e., to the pressure of the saturated vapor that is present already), so that p represents the same quantity in both formulas. Substitution into the 2d equation yields

$$\frac{A_2}{J} = D - \frac{D}{T} \cdot \frac{1}{\frac{1}{p}\frac{dp}{dt}}.$$

Obviously, we have now just a function of temperature on the right-hand side.

Substituting the corresponding functions for ether, for example, one obtains

$$\frac{A_2}{J} = 82.9 - 0.0990 \text{ t } <+> - 0.00265t^2$$

$$t' = 197°$$

[Note in margin:] Not correct. Errors in the constants.

Let t' be the particular value of the temperature at which $\frac{A_2}{J}$ becomes 0. This is the value of the temperature at which the transition from the liquid to the gaseous state is no longer associated with an expenditure of energy. The cohesive force is here exactly cancelled by the actions of thermal factors. In the case of saturated vapor there will thus arise an indifferent equilibrium between the liquid and gaseous states. [Note in margin:] Obscure point.

COURSE IN PHYSICS BY PROF. WEBER. ALBERT EINSTEIN VI A.

On the liquefaction of gases.

The cohesive force varies with the temperature. If we have a gas that we compress at a certain temperature to such a degree that the molecular attraction forces can again become effective, then this gas must liquefy until the portion that remains in the gaseous form reaches a lower density. Thus we obtain a saturated vapor.

The compression of the gas can occur in the following way.

[Fig.]

gas to be investigated

This apparatus permits us to subject the gas enclosed in the tube to an arbitrary pressure by pumping arbitrary amounts of water into the external vessel and thus changing the volume of the gas. This apparatus permits the arbitrary raising of the pressure if the external vessel is sufficiently resistant, because the internal tube does not have to withstand any pressure.

We now set up the following arrangement.

[Fig.]

pump
air
SO_2

This apparatus shall [serve] for the determination of the relation between temp. and liquefaction densities. If water is pumped in, the two mercury levels rise at first in the same way, proof that the equation of state is still satisfied. At a temperature of 20° this takes place at up to 4 atmospheres. If additional water is pumped in, the level of the tube on the left remains steady, while that of the tube on the right has risen considerably. However, the left tube has only the function of indicating the pressure in the vessel.

If the condensation on the left has stopped despite continued pumping, this proves that the pressure remained the same, since at these pressures the air still follows Mariotte's law almost strictly. If the volume on the left [corrected to "right"] decreased significantly without an increase in pressure having taken place, we conclude that SO_2 converted to saturated vapor. This is also indicated by the tube becoming more cloudy, which is to be attributed to the deposition of small liquid droplets. The tube on the left permits us to determine the liquefaction pressure. Assuming that Mariotte's law is rigorously obeyed.

$$\frac{V_{p_0}}{V_p} = \frac{p}{P}$$

$$p = P \cdot \frac{V_{p_0}}{V_p} = P \cdot \frac{\ell_0}{\ell}.$$

Here in CGS. $P = 76 \cdot 13.596\ g$ |gr cm.
If we note the heights in the tube on the right at the moment of conversion, we obtain the vapor density corresponding to the observed temperature.
For experiments at lower temperatures one surrounds the whole thing with a cooling mixture down to -22° (27 [parts] ice 17 [parts] salt). To produce even lower temperatures one uses liquid carbon dioxide.

carbon diox[ide] gas [Fig.]

Preparation of the same according to the principle shown in the diagram.

At 0° carbon dioxide liquefies at 38.5 atmospheres.

Relation between p and t for carbon dioxide

t	p
21	60
13	47
0	38
-12	27
-29	16
-51	7
-78	1

If liquid carbon dioxide is allowed to evaporate at low pressure and a large surface area, then this happens so vehemently that the heat required for the evaporation, which is extracted from the liquid carbon dioxide still present, is sufficient to cool the latter so much that it solidifies. It forms an amorphous white substance.

Faraday investigated a great number of gases in this way and found that by using elevated pressures he could liquefy all of them except for

$$\begin{array}{ccc} H_2 & N_2 & O_2 \\ CH_4 & NO & CO. \end{array}$$

Later 1854-55 Natterer carried out further experiments, he tried to liquefy air at normal temperatures.

Since he wanted to proceed to pressures of up to 3000 atm., he introduced a direct measurement of the pressure.

His experimental setup was as follows.

pump [Fig.]

Natterer determined also the relation between pressure and density.

$$fp = mg \quad 16\cdot 11 = \underline{m} \; 176 \; \underline{g}$$

$$p = \frac{m\cdot 176\; g}{f} \text{ (in absolute units } |\text{gr cm}|\text{).}$$

We have, then

$$P = 76\not{0}0.013/596g.$$

. . . .

The number of atmospheres $= \frac{p}{P} = \frac{m\cdot 176\; g}{760\cdot 0.013.596g\cdot f} =$

If one takes as unit

$$\frac{p}{P} = \frac{m\cdot 176}{760\cdot 0.013596\cdot f\cdot 1\not{0}\not{0}} = \frac{m}{0.46} \quad \text{(at the } f \text{ used).}$$

If the mass of the enclosed gas in each test is known, then a series of tests yields a relation between ρ and p for the gas investigated. The amount of gas contained is determined in the manner indicated.

If the vessel has to be filled n times, then the total mass of the gas contained in the large vessel before is

$$M = m\cdot n$$
$$m = v\cdot \rho.$$
$$\rho = p\cdot \frac{\rho_{n_0}}{p_n}\cdot \frac{1}{(1+\alpha t)}.$$

Consequently,

$$m = <v\cdot p\,\frac{p_n}{p_{n_0}}\,(1+\alpha t)> \qquad v\cdot \frac{p}{p_n}\cdot \rho_{n_0}\left(\frac{1}{1+\alpha t}\right).$$

Here p is the prevailing atmospheric pres. If M is known, one obtains

$$\rho_{\text{(in t[he] l[arge] vol[ume])}} \frac{M}{V}.$$

The experiments showed that $p \cdot V$ or $\frac{p}{\rho}$ were no longer constant as required by Mariotte's law, but rose between 1 and 3000 atm. in the ratio 26:100. The reason for the failure of Natterer's liquefaction experiments was elucidated by Andrews in 61-65. He set himself the task of thoroughly studying *one* gas that was relatively easy to investigate with respect to its properties at higher pressures and varying temperatures, in order to draw from this conclusions about permanent gases. He chose CO_2, which liquefies at relatively accessible conditions, and investigated the relation between v and p at various temperatures. He used the following setup.

[Fig.]

air temp. $\qquad$ CO_2 temp.

Again, the tube on the left serves for the measurement of pressure. The results he found were essentially as follows.

[Fig.]

The experiments show <show> that the ratio of the volume at which the gas starts turning into saturated vapor and the volume at which the substance manifests liquid characteristics is a quantity that decreases with increasing temperature and tends toward 1. Once this limit is reached, the transition through the state of saturated vapor does no longer take place and one can no longer talk about a separation between the gaseous and the liquid form. The temperature at which this sets in is +31°, i.e., the same one we already designated earlier as critical temperature. Thus, the failure of Natterer's experiments was due to his investigating the air at a temperature higher than the critical temperature. He himself carried out further experiments with permanent gases but did not reach their inferred critical temperature. He used liquid carbon dioxide as a refrigerant (down to -78°).

Thus, one had to work at lower temperatures. In 1877 Cailletet carried out the following investigation:

He made use of the cooling that occurs during adiabatic expansion.

We have

$$\left(\frac{p_1}{p}\right)^{\frac{\kappa-1}{\kappa}} = \frac{T_1}{T}.$$

If a gas compressed <near> 0° is allowed to expand adiabatically, this formula yields the following temperatures at different vapor pressures.

[145]

$\frac{p}{P}$	t
50	-184
100	-200
200	-213
300	-220
400	-224
500	-227

The following apparatus was used for the cooling itself. (For air κ = 1.405). For less heavy gases too.

compressed gas
metal capillary tube [Fig.]
water pump

One suddenly opens a valve so that the gas suddenly expands and water [flows] from the vessel. The whole process shall last 1-2 seconds. Now the following will happen:

If the gas has been expanding for some time, its temperature will have significantly decreased because of the work done by it. However, the prevailing pressure is still quite considerable so that one would expect a liquefaction.

Cailletet investigated in this way the gases found by Faraday and Natterer to be constant, and found that a turbidity had set in in the gas which almost instantaneously disappeared again. Proof that liquefaction has set in, because the turbidity <was due> must have been due to small particles of liquid. However, owing to the relatively fast heat inflow, at the prevailing large temperature difference, gasification had almost instantaneously set in again. In this way it has been established that it is possible to liquefy all gases, only about hydrogen were the observers not quite sure.

Pictet's method.

This [method] meant to produce lastingly low temperatures. He used the heat of evaporation.

pressure pump suction pump gas to be tested

[Fig.]

suction pump cooling vessel

In these experiments O was produced in a retort by heating $KClO_3$, which simultaneously gave rise to the pressure needed for the liquefaction.

The critical pressure is the upper limiting value the pressure of the saturated vapor can have, the pressure of the saturated vapor at the critical temperature.

Electricity.

To find the fundamental laws governing the behavior of electric quantities, we must get to know one of the ways that lead to the appearance of electric forces.

Electricity is generated by the mere contact of any two substances. (Friction electricity is to be attributed to this, since mere contact suffices.) We need an apparatus to measure small forces. For this we use a bifilarly suspended mass.

[Fig.]

We determine the torque

$$\text{Torque} = \frac{d\mathcal{E}}{d\nu} = \frac{1}{\ell}\,\frac{d\left[mg\cdot\langle\ell\rangle 1 - \sqrt{\langle\ell^{2}\rangle 1^{2} - \dfrac{\left(2a\sin\frac{\nu}{2}\right)^{2}}{\ell^{2}}}\right]}{d\nu}$$

$$\left\langle mg\cdot\frac{2a\sin\frac{\nu}{2}\cos\frac{\nu}{2}\cdot d\nu}{\sqrt{\ell^{2}-\left\{2a\sin\frac{\nu}{2}\right\}^{2}}} = mg\;2a\sin\frac{\nu}{2}\cos\frac{\nu}{2}\left\{\ell + 2a\sin\frac{\nu}{2}\right\}\right\rangle$$

$$\frac{1}{\ell}\,\frac{d\left[mg\cdot 2a^{2}\sin^{2}\frac{\nu}{2}\right]}{d\nu} = \frac{1}{\ell}\,mg\;2a^{2}\cdot 2\sin\frac{\nu}{2}\cos\frac{\nu}{2}$$

$$= mg\langle 2\rangle\,\frac{a^{2}\sin\nu}{\ell}.$$

If at a distance α a force K acts upon the rod of mass m perpendicularly to the mid-vertical, then the system is in equilibrium if the following condition is satisfied:

$$K\alpha = \frac{mga^{2}}{\ell}\sin\nu.$$

This is a means for measuring K. If the forces are very small, instead of bifilar suspension one can also use the torque of the thread. This is $\nu \cdot \frac{\pi}{2} \frac{c\rho^4}{\ell} = \theta\nu$.

The concept of electric masses was developed next, and different signs were introduced for them to facilitate the expression of the existing laws.

Using contact electricity, it has been proved that electricities of the opposite kind attract each other, and those of the same kind repel each other.

[Fig.]

mica leaf with gold   ebonite   gold-covered glass plate

Coulomb's law.

Coulomb started out from the following Ansatz:

$$f = \frac{\alpha e_1 e_2}{r^n}.$$

It had to be determined experimentally and shown that f is directly proportional to $(e_1 \cdot e_2)$.

[Fig.]

Coulomb first imparted a charge of $\pm e$ to the little sphere A at rest and then by direct contact passed on half of the charge to the equal-sized little sphere B. The little sphere B then assumed a certain equilibrium position.

$$\frac{\alpha e_1 e_2}{r^n} \cdot \cos \frac{x}{2} \cdot \ell = \theta x$$

where θ has already been determined previously. If the screw was turned clockwise by the $\measuredangle\ \alpha$ and B took up a new position x', then

$$\frac{\alpha e_1 e_2}{r'^n} \cdot \cos \frac{x'}{2} \cdot \ell = \theta(x' + \alpha).$$

By division

$$\left(\frac{r'}{r}\right)^n \frac{\cos \frac{x}{2}}{\cos \frac{x'}{2}} = \frac{x}{x' + \alpha}.$$

Since all quantities except n are known, it is possible to determine the latter. If one were to change the arrangement by discharging the sphere A and bringing it into contact with B anew so that the charges in both become $\frac{e}{2}$, one would obtain a 4 times smaller force, proof that the force is proportional to $e \cdot e'$.

Since the units of e and $-e$ are yet to be chosen, one can choose $\alpha = 1$. It then holds generally that if opposite signs are assigned to the masses, the general expression of Coulombs's results is

$$f = \frac{e_1 \cdot e_2}{r^2}$$

where the repelling force is taken as positive.

To determine the force exerted by some arbitrary mass distribution on an <outside> electric mass, one makes use of a function of the coordinates, the potential P, which depends on the given mass distribution and the coordinates $x\ y\ z$ of a point considered. We choose by definition

$$P = \sum \frac{e_n}{r_n}.$$

[Fig.]

Now we have to find the main properties of the pot[ential].

$$\frac{\partial P}{\partial x} = -\sum \frac{e_n}{r_n^2} \cdot \frac{(x - x_n)}{r_n} = -\sum X_n = -X.$$

We have then

$$\left.\begin{aligned} X &= -\frac{\partial P}{\partial x} \\ Y &= -\frac{\partial P}{\partial y} \\ Z &= -\frac{\partial P}{\partial z} \end{aligned}\right|$$

Since the <direction> position of the chosen system of axes is completely arbitrary and any straight line can be taken as the x-axis, it follows for the force component S on the straight line s

$$S = -\frac{\partial P}{\partial s}.$$

$-\frac{\partial P}{\partial s}$ represents the gradient of the potential in the direction s.

$$\overset{\bullet}{P} \overline{\quad ds \quad} \overset{\bullet}{P} + \frac{\partial P}{\partial s}\, ds = P' \qquad -\frac{\partial P}{\partial s} = \frac{P - P'}{ds}.$$

Thus one has the law in general.
The force acting on the mass +1 at an arbitrary point equals the [negative] gradient of the potential in the direction indicated.

We now imagine that the point of mass +1 is moving freely under the influence of the electrical mass forces and we will investigate the work it performs.

[Fig.]

$$dA = Sds = -\frac{\partial P}{\partial s} \cdot ds = -dP$$

$$\int_{P_1}^{P_2} dA = - \int_{P_1}^{P_2} d<p>P$$

$$A = P_1 - P_2 .$$

Thus, for electrical forces, based on Coulomb's law, the work performed is independent of the path of integration.

If P_2 is removed to infinity, then P_2 becomes = 0 and the equation reduces to $A = P_1$. I.e.: The potential for a point (xyz) equals the work the electrostatic forces would perform if the positive unit mass were moved from that point to infinity. Or, it is also the work that would have to be spent to bring the mass +1 from infinity to the position under consideration.

Given is a body for which it is assumed that electricity can freely move in its interior and on its surface. A certain amount of electricity shall now be brought to the body from the outside, say in A.

[Fig.]

The distribution will possess a certain potential function. We now consider the arbitrary point P. In this point the potential function will have a certain value. A point P' farther outside will have a smaller potential, consequently $\frac{\partial P}{\partial s}$ will be different from 0, hence also the quantity of electricity at point P will be acted upon by a force which will seek to move it outward. The electric mass will thus be dispersed and will achieve the state of equilibrium only after P had achieved the same value throughout the body.

Thus, P is a quantity associated with the body which is independent of position and is called its electrostatic potential.

[Fig.]

If we consider the points in the body's surroundings, they possess a smaller potential. Since the potential is a continuous function of position, there exists an infinite number of equipotential surfaces that surround the body. The surface of the body itself is such a surface since its potential is everywhere P.

If the equipotential surfaces are known, it is possible to derive from them the direction and magnitude of the acting forces for each point in space with the help of the following propositions.

1) The force acting in an arb[itrary] point on the mass +1 is ⊥ to the equipotential surface. If the force is resolved into 2 components, one perpendicular to the equipotential surface, the other lying in it, then, according to the above, the latter

component $= -\frac{\partial P}{\partial s} = 0$ because according to assumption P [is] constant on the surface.

[Fig.]

2) The magnitude of the force is $= \frac{P_1 - P_2}{\Delta n}$ since this quantity represents the [negative] gradient.

If one plots the direction of the force for every point of space, one obtains the body's total system of lines of force as the system of the orthogonal trajectories of the equipotential surfaces. Since the surface of the body is an equipotential surface, the direction of the force in each of its points is the direction of the normal.

[Figure caption]: Plot of values of potential for the points of a straight line that traverses an electrically charged conductor

[Fig.]

$\frac{\partial V}{\partial \ell}$ experiences in this case a discontinuity on the surface.

Measurement of the potential of a conductor. Thomson's quadrant electrometer.

ground [Fig.]

A light aluminum foil that is kept constantly at a certain potential value is bifilary suspended over metal quadrants that are conductively connected in the way indicated. The potential of the earth is put = 0. If there exists a potential difference between the quadrant pairs, then the foil of potential P_0 will experience a torque

$$\text{Torque} = c\ (P_1 - P_2)\ [P_0 - \tfrac{1}{2}(P_1 + P_2)].$$

We now bring circuit 2 in the apparatus in contact with the ground and make P_0 very large relative to P_1, then we have approximately

$$\text{Torque} = cP_1P_0.$$

If the foil is in equilibrium, then besides this torque there acts another one of the same magnitude which is produced by the bifilar suspension.

$$\text{Torque} = Mg\,\frac{ab}{L}\sin x.$$

Thus we have

$$\sin x = \left| \frac{cP_0L}{Mgab} \right| P_1.$$

This equation provides the means for increasing the deflection x when P_1 is small, by a suitable choice of constants. It now remains to see to it that the value of P_0 be always the same. This is achieved by the foll[owing] arrangement.

potential of ground
fixed hinge [Fig.]
weight
insulation

The mobile part of the Thomson instrument 300 mgr including the mirror.

The electrometer is used to demonstrate conductivities of various substances.

solution of H_2SO_4 [Fig.]

Volta's fundamental experiment on the galvanic element.

Using the electrometer, one shows that the zinc rod takes on a negative potential. If it is replaced by a copper rod, then the latter also takes on a negative potential, though of a lesser magnitude. If one has simultaneously two rods in the liquid, one of each metal, then a potential difference exists between them, which forms anew no matter how often one removes the charge electricity. If the two metals are connected by a conductor, then, due to the constant potential difference and the potential gradient at each point of the conductor associated with it, a constant electric current is formed whose positive pole is the <carbon> copper and whose negative pole is the zinc.

Distribution of the electric charge of a conductor at electric equilibrium.

We seek to ascertain the action of all the masses on the mass 1 in a point of the surface in the direction of its normal.

[Fig.]

$$\frac{e_1}{r_1^2} \cos \alpha_1 \; df = e_1 \cdot d\sigma = n_1 \cdot df .$$
$$\Sigma \; e_1 \; \underline{d\sigma} = \Sigma \; n_1 \; df$$

$$4\pi e_1 = \Sigma \; n_1 \; df$$
$$4\pi e_2 = \Sigma \; n_2 \; df$$

$$4\pi \; \Sigma \; e = \Sigma \; n \; df = \int n \; dF .$$

For the external masses E:

[Fig.] $$df \frac{E_1}{R'^2} \cos A' = N_1' df' = E_1 \, dT$$

$$df \frac{E_1}{R''^2} \cos A'' = N_1'' df'' = - E_1 \, dT$$

$$N_1' df' + N_1'' df'' = 0.$$

Summing over the surface, one obtains

$$\int N_1 df = 0.$$

Summing over the individual masses as well, one has

$$\int Ndf = \int (N_1 + N_2 \ldots) df = 0.$$

If one denotes by $\mathcal{N}$ the components of the normal which is composed of n and N, one obtains by adding the two results:

$$\int \mathcal{N} \, df = 4\pi \, \Sigma e.$$

If one has an electric conductor that is in the state of equilibrium and if one marks off an arbitrary closed surface in its interior, then $\int \mathcal{N} df$ must vanish because equilibrium exists only if no forces act in the interior on the masses that might be present there. Thus, for every surface

$$\Sigma e = 0$$

which is only possible if e is everywhere = 0 or if everywhere $+e$ and $-e$ are equal.

Thus, if one assumes that the conductor is adjacent to a nonconductor, which cannot take up the electric quantities by conduction, then one is left with the notion that the electric masses accumulate on the surface of the conductor in the form of a fine layer, so that the density of this layer depends on the strength of the charge and the shape of the surface at the location in question.

This leads to the definition of the surface density σ of electricity, the mass which belongs to the unit area of the conductor surface at a specific location. It is usually very difficult to determine the equilibrium distribution of σ for a charge of a specific magnitude and for a specific body shape.

On the other hand, if the surface distribution of σ is presumed to be known, then the electrostatic force acting toward the surface can easily be calculated for every spot of the conductor.

If the Gauss law is applied to the rectangular prism in the accompanying diagram, one obtains

[Fig.] $$\mathcal{N} \, df = 4\pi\sigma \, df$$
$$K = \mathcal{N} = 4\pi\sigma$$

The distribution of σ at different E's obeys the rule that if we add E_1 E_2 ... to the body and if σ_1 σ_2 ... are the densities as functions of the position, then $E_1 : E_2 : E_3 = \sigma_1 : \sigma_2 : \sigma_3$.

Because if one imagines that the distribution σ_1 which corresponds to E_1 has been established and that σ_1 has been replaced everywhere by $\mu\sigma_1$, then one recognizes that equilibrium must again obtain for each point of the surface, since it existed earlier. However, since the distribution of a fixed quantity is unique, this distribution is the only one possible for μE. Hence σ changes proportionally to E.

The relation between E and P for a conductor.

The relations

$$P = \int \frac{\sigma df}{r}$$

$$E = \int \sigma df$$

[ ].

If E increases to μE, then, according to the foregoing, σ increases to $\mu\sigma$ and consequently, P to μP. Thus, it holds that

$$E = CP.$$

C is a constant which is called the capacity of the conductor. This capacity equals in the above-mentioned system of measurements = $L^{<-1>}$ as can be realized from the defining equation. The mathematical determination of C is usually very difficult, because the distribution of σ [is] very difficult to determine. C is usually determined experimentally. According to the theory developed, C is independent of the material of the conductor and depends only on the shape of the boundary.

Experimental determination.

Since the experimental determination of C is of the comparative kind, it is necessary to calculate first the capacity of a specific body, e.g., a metal sphere.

[Fig.]

$P = \frac{E}{R} =$, because the distribution of σ [is] constant

$C = \frac{E}{P} = R.$

The capacity of a thin circular disc

$$C = \frac{2R}{\pi}.$$

First one has to determine the capacity of the electrometer used.

We impart a certain potential P to the quadrant pair. The instrument produces a deflection so that

$\sin X = A \cdot P$, where A is a constant that depends on the apparatus of electrification of the mobile part of the instrument. [Probably part of this sentence should have been crossed out.]

By means of a noncapacitive wire we now connect the quadrant pair with an insulated metal sphere and wait for equilibrium to be established. The system is then at the potential p, so that

$$\sin x = Ap.$$

Consequently, one has $\dfrac{\sin X}{\sin x} = \dfrac{P}{p}$.

However, the quantity of electricity is of the same magnitude in both tests. If C is called the capacity of the instrument, one obtains

$E = C \cdot P = Cp + Rp$, where R [is] the radius of the sphere.

$$\frac{C(P - p)}{p} = R = C\left\{\frac{P}{p} - 1\right\} = C\left\{\frac{\sin X}{\sin x} - 1\right\}$$

$$\boxed{C_i = \frac{R}{\dfrac{\sin X}{\sin x} - 1}}$$

After the capacity of the electrometer has been determined in this way, we determine the capacity of another conductor by the same procedure, except that now the instrument takes the place of the sphere and the body to be tested takes the place of the instrument.

We denote the capacities of the instr. and the object by C_i and C [respectively].

$$E = C_i\ P = C_i p + Cp$$

$$\frac{C_i(P - p)}{p} = C$$

$$\boxed{C = \left\{\frac{\sin X}{\sin x} - 1\right\} C_i .}$$

Induction.

If an electric quantity of a certain sign is brought near an electrically neutral conductor, then an opposite-signed electric quantity will accumulate on the sides of the conductor that face the approaching quantity, and a quantity of the same sign on the opposite side (induction). The process is explained by the forces exerted according to Coulomb's law by the approaching masses on the masses of

[164]

the opposite sign which are evenly distributed in the conductor. This distribution goes on until the potential in the interior of the conductor has everywhere become constant.

The determination of the mass distribution in specific cases is again very difficult. We will here consider a few simple cases only.

[Fig.]

1. A certain quantity e is brought into the interior of a hollow metal conductor. We wish to determine the quantity E accumulated on the inner surface.

If one demarcates a closed surface that is located entirely inside the conductor, then for it $\Sigma\ N df = 0$.

Consequently, the law of Gauss will yield

$$\Sigma\ e = 0 \quad \text{or} \quad E = -e.$$

The quantity of electricity that is accumulated on the inner surface considered is equal to that in the interior of the cavity, but has the opposite sign.

2) Condenser

ground [Fig.]

We set up the arrangement shown in the diagram. The metal plate on the right shall have the potential P, and that on the left shall be connected with the ground, i.e., shall have the potential P_e. We want to determine the electric masses accumulated on the inner surfaces of the metal plates.

The calc[ulation] of the potential difference

$$P - P_e.$$

The potential gradient

$$\frac{P - P_e}{a} = K.$$

Consequently, in accordance with Gauss's theorem, the quantity σ on the unit surface area is

$$\sigma = \frac{P - P_e}{4\pi a}.$$

Thus, the total electrical layer of the plate on the right is

$$\underline{\mathcal{E}} = \frac{P - P_e}{4\pi a}\underline{F}.$$

All this applies equally to the plate on the left, except that the potential difference will change in the direction of the normal. Thus, σ and $\mathcal{E}$ are of the same magnitude and of opposite sign.

The quantity $\frac{F}{4\pi a}$ is called the capacity of the condenser analogously to the previous case. The coatings of electrical masses of the other surfaces are so insignificant relative to these two that they can be neglected.

Two such condenser plates can be used for the generation of a homogeneous field of force.

We investigate the value of the potential that the two layers exert on an outside point.

$$dp = \frac{\sigma df}{r_+} - \frac{\sigma df}{r_-}$$

[Fig.]

$$r_+^2 = r^2 + \frac{a^2}{4} - ar \cos \varphi$$

$$= r^2 - ar \cos \varphi$$

$$r_+ = r(1 - \frac{a}{2r} \cos \varphi) = r + \frac{a}{2} \cos \varphi .$$

Analogously,

$$r_- = r - \frac{a}{2} \cos \varphi$$

$$dp = \sigma df \left[\frac{1}{r + \frac{a}{2} \cos \varphi} - \frac{1}{r - \frac{a}{2} \cos \varphi} \right] = \frac{\sigma \; df \; a \cos \varphi}{r^2}.$$

$$P = \int \frac{\sigma \; df \; a \cos \varphi}{r^2} = a\sigma \int \frac{df \cos \varphi}{r^2} = a\sigma \int d\kappa = \qquad .$$

$\int dK$ denotes the magnitude of the surface area of the unit sphere cut out by the cone that can be drawn from P towards the boundary of the plates.

Since a is a small quantity, the value of this potential becomes very low for all points outside, hence also its gradient, hence also the force exerted outward by the layers on the condenser plates.

Measurement of the charge E of the condenser.

Let one condenser plate be brought to the potential P while the other one is conductively connected from above with the ground. Let us determine the charge E of the first plate.

[Fig.] ground

We first charge the plate on the right, then remove the plate on the left, and connect the plate on the right, whose capacity shall be $= c$, with the electrometer.

A state of equilibrium is then reached in the latter, which permits the determination of the potential p of the present system.

$$\mathcal{E} = p(C_i + c)$$

$$p = A \sin x \qquad \langle A = C.P_0\rangle$$

(A is dependent on the instrument and indirectly proportional to P_0).

One thus has

$$\mathcal{E} = A(C_i + c) \sin x$$

[The following 5 equations are in margin:]

$$PC_i = p(C_i + c)$$

$$P = A \sin X$$

$$p = A \sin x$$

$$C_i \frac{[P - p]}{p} = c$$

$$C_i \left[\frac{\sin X}{\sin x} - 1\right] = c$$

c is to be determined *separately* according to the above method.

The Leyden flask is a completely analogous type of condenser.

Based on the above principle, it is easy to construct a condenser of large capacity using the following scheme.

[Fig.]

If one has n plate pairs, then the capacity is

$$(2n - 1) \frac{F}{4\pi a}$$

$$\mathcal{E} = 2n - 1 \frac{P - P_e}{4\pi a} F.$$

Capacity of the cable.

We examine the distribution of the potential in the dielectric substance.

[Fig.]

metal
insulation
core
metal

We apply the Gauss law to an element

$$2r\pi\ dx\cdot\frac{dp}{dr} + 2(r + dr)\ \pi\ dx\cdot\left\{-\frac{dp}{dr} - \frac{d^2p}{dr^2}\,dr\right\} = 0.$$

Neglecting the t[erms] of higher order, one obtains

$$\frac{d^2p}{dr^2}\,r + \frac{dp}{dr} = 0$$

$$\frac{\partial}{\partial r}\left[r\ \frac{dp}{dr}\right] = 0$$

$$r\ \frac{dp}{dr} = a$$

$$\frac{dp}{a} = \frac{dr}{r}$$

$$\frac{p}{a} = \lg\ (\frac{r}{c});\qquad p = a\ \lg r + b.$$

For the boundary one must have

$$\lg r_2 \quad\Big|\quad P = a\ \lg r_1 + b \qquad a = -\ \frac{P - P_e}{\lg \frac{r_2}{r_1}}$$

$$\lg r_1 \quad\Big|\quad P_e = a\ \lg r_2 + b \qquad b = \frac{P\ \lg r_2 - P_e \lg r_1}{\lg r_2 - \lg r_1}.$$

Surface density of el[ectricity] on the surface of the core:

$$\sigma = -\ \frac{1}{4\pi}\ \frac{\partial P}{\partial r} = \frac{1}{r}\cdot\frac{P - P_e}{\lg \frac{r_2}{r_1}}\cdot + \frac{1}{4\pi}$$

$$\sigma_1 = \frac{1}{r_1}\boxed{\frac{1}{4\pi}\ \frac{P - P_e}{\lg \frac{r_2}{r_1}}}$$

$$\sigma_2 = \frac{1}{r_2}\boxed{\frac{1}{4\pi}\ \frac{P - P_e}{\lg \frac{r_2}{r_1}}}$$

$$\varepsilon_1 = \varepsilon_2 = \frac{1}{2}\ \frac{P - P_e}{\lg(\frac{r_2}{r_1})}$$ per unit length.

Capacity: $\frac{1}{2\lg(\frac{r_2}{r_1})}$.

Dielectric constant.

If one places some substance between the two surface layers of a condenser and measures the true charge E at a fixed potential difference in the above described manner, then E depends on the nature of the substance and differs from the value given above by a constant δ characteristic of the substance.

$$\varepsilon_{\text{subst.}} = \varepsilon_{\text{air}} \cdot \delta_{\text{subst.}}$$

δ is called the dielectric constant. From this it follows, dividing by $P - P_e$, that

$$C_{\text{subst.}} = C_{\text{air}} \cdot \delta .$$

δ is always larger than 1.

Subst.	δ
Air	1.00
Paraffin	1.92 } 2.47
Ebonite	2.6 } 3.3
Mica	4.6 } 5.2

The values found above for the capacity of condensers and cables must thus be multiplied by the δ of the substance involved.

A condenser of very large capacity can be obtained by combining mica and tinfoil.

Capacity insulation of mica

$n \frac{F}{4\pi a} \delta$ [Fig.]

where n is the number of mica foils.

Using such an arrangement and a galvanometer, we demonstrate the magnetic effect of the current which appears when the charge disappears. Explanation of the appearance of the dielectric constant by assuming the electric polarization of the molecules.

[Fig.]

The interposition of the molecules causes a reduction of P when $\mathcal{E}$ constant. Thus an increase of capacity.

Electric currents.

Let there be given a Voltaic cell; with the help of the electrometer one demonstrates that the two poles constantly possess a definite potential difference.

ground [Fig.]

If we connect the Cu electrode with the electrometer, and the Zn el[ectrode] with the ground,/then the electrometer shows a constant positive deflection, and when the cell is reversed a constant negative deflection of the same magnitude. One shows that the shape of the electrode does not affect the magnitude of this potential difference.

We now connect the two electrodes by a conductor.

[Fig.]

By a linear conductor we understand a conductor whose cross-sectional dimensions we neglect, and we count the arc lengths of the conductor starting from A, so that a point $\mathcal{P}$ is defined by its arc length x. However, the potential difference between the ends of the conductor is a constant magnitude. But to the potentials of the electrodes there correspond mass layers on the electrodes, and to the latter, values of the potential in the entire surrounding space which are a function of position. On each continuous curve drawn from A to B one will obtain a continuous function for the potential.

$$f(x) = P \quad f(0) = P_1 \quad f(\ell) = P_2 .$$

The moment the metal circuit has been set up, electric masses will start to move in the wire in the direction of the acting force and these in turn will change the space function of the potential until a stationary state has been established. The distribution of the potential in the conductor depends on the law governing the motion of the electric masses in the conductor. This we assume hypothetically.

We denote by i the quantity of electric masses passing through a particular cross section of the conductor during time 1 and formulate the hypothesis:

$$i = \kappa \cdot q \cdot \left(- \frac{\partial p}{\partial x}\right)$$

where q denotes the cross section of the conductor, p the potential [and] κ a characteristic of the material which is also especially dependent on the temperature.

Based on this hypothesis we consider the flow in the conductor.

[Fig.]

the quantity flowing in during time dt $\quad (\kappa\cdot q\cdot - \frac{\partial p}{\partial x})\, dt$

the quantity flowing out " " dt $\quad \left[\kappa\cdot q\cdot - \left\{\frac{\partial p}{\partial x} + \frac{\partial^2 p}{\partial x^2}\, dx\right\}\right] dt.$

Excess of the inflowing quantity $\quad \kappa q \frac{\partial^2 p}{\partial x^2}\, dx\, dt.$

This excess can also be expressed as cap.$\cdot dx\cdot\frac{\partial p}{\partial t}\cdot dt.$

Since the two quantities must be equal, it follows that

$$\boxed{\frac{\text{cap}}{\kappa\cdot q}\cdot\frac{\partial p}{\partial t} = \frac{\partial^2 p}{\partial x^2}.}$$

This is the differential equation of linear electric currents. However, since in our case of a good electric conductor κ is very large and $\frac{\text{cap}}{q}$ is always a small quantity, it follows that the left-hand side can be neglected without a noticeable error. It follows therefore with complete generality, thus even when P_1 and P_2 are functions of time:

$$\frac{\partial^2 p}{\partial x^2} = 0$$

$$< \frac{\partial^2 p}{\partial x} > p = ax + b$$

In our case we will then have

$$p = P_1 - \frac{P_1 - P_{<0>2}}{\ell}\cdot x.$$

Thus, if the hypothesis about the electrical current is correct, then for a homogeneous wire of constant cross section, the potential must satisfy this condition. Measurement with the electrometer showed coplete agreement.

It follows further

$$i = \kappa\cdot q\cdot - \frac{\partial p}{\partial x} = \kappa q\cdot\frac{P_1 - P_0}{\ell} = \frac{P_1 - P_{<0>2}}{W} \qquad \text{where } W = \frac{\ell}{\kappa\cdot q}.$$

The quantity $\frac{1}{\kappa} = \omega$ is also called the specific resistance of a substance. Thus i is independent of x, and thus a constant.

If one multiplies by W, one obtains the law found by Ohm in another way

$$iW = P_1 - P_2.$$

Here W is the conductor's resistance $= \frac{\ell}{\kappa \cdot q} = \omega \frac{\ell}{q}$. It should be noted that i depends only on $P_1 - P_2$ but not on time.

[Fig.]

Let us have a combination of 2 conductors of lengths ℓ_1 and ℓ_2 and of the same constant cross section Q. We are seeking the distribution of the potential and the current intensity.

$$i_1 = \frac{P_1 - p'}{\ell_1} \cdot \kappa_1 q$$

$$i_2 = \frac{p' - P_2}{\ell_2} \cdot \kappa_2 q.$$

But we must have $i_1 = i_2$.

$$\frac{P_1 - p'}{\ell_1} \cdot \kappa_1 = \frac{p' - P_2}{\ell_2} \cdot \kappa_2$$

$$(P_1 - p')\, \kappa_1 \cdot \ell_2 = (p' - P_2)\, \kappa_2 \ell_1$$

$$\frac{P_1 \kappa_1 \ell_2 + P_2 \kappa_2 \ell_1}{\kappa_1 \ell_2 + \kappa_2 \ell_1} = p'$$

The [potential] gradients are inversely proportional to the conductivities for equal cross sections.

Current and potential distribution in a Voltaic element.

If one considers an electric mass that is moving in an electrical circuit, it passes a location of discontinuous change of potential at every bounding surface. We consider the potential to be constant within the electrodes. To explain the analogy between this kind of current and linear current, we plot lines of direction of current in the space, so that the space gets divided into a number of tubes. Let the quantity flowing through the filament (1) be M_1

[Fig.]

$$M_1 = \kappa' dq_1 \left(-\frac{dp}{dx_1}\right)$$

$$di = \kappa' dq_1 \left(-\frac{dp}{dx_1}\right)$$

$$\frac{1}{\kappa'} \int_0^{\ell_1} \frac{di}{dq_1}\, dx = \int_0^{\ell_1} - dp$$

$$P_4 - P_5 = \frac{di_1}{\kappa'} \int_0^{\ell_1} \frac{dx}{dq_1}.$$

[The following 3 equations are in margin:]

$$\frac{di_1}{(P_4 - P_5)\kappa'} = \frac{1}{\int_0^{\ell_1} \frac{dp}{dq_1} dx}$$

$$\frac{I}{(P_4 - P_5)\kappa'} = \sum \frac{1}{\int_0^{\ell_n} \frac{dp}{dq_n} dx_n}$$

$$W = \frac{1}{\kappa'} \cdot \langle \sum \rangle \frac{1}{\sum \frac{1}{\int \frac{dp}{dq_n} dx_n}} .$$

This holds true for every filament. [Note in margin:] Truly legerdemain! Assumption of that which has to be proven.

$$P_4 - P_5 = \frac{di_2}{\kappa'} \int_0^{\ell_2} \frac{dx}{dq_2}$$

$$di_n = \frac{\not{\kappa'}}{\not{P_4} - \not{P_5}} \cdot \frac{P_4 - P_5}{\sum \int_0^{\ell_n} \frac{dx}{\kappa' dq_n}}$$

If one now defines $\sum \int_0^{\ell_n} \frac{dx}{x' dq_n} = W \rangle$

one ob[tains]

$$i = \frac{P_4 - P_5}{W'} .$$

Thus, an analogous law is valid here

$$iW = P_1 - P_2$$

$$iW' = P_4 - P_5$$

$$i(W + W') = P_1 - P_2 + P_4 - P_5$$

$$= \underbrace{(P_1 - P_{\langle\langle 5 \rangle 4 \rangle 5}) + (P_{\langle\langle 4 \rangle 3 \rangle 4} - P_{\langle\langle 3 \rangle 2 \rangle 3}) + (P_3 - P_2)}_{\text{the sum of electromotive forces}}$$

$$i(W + W') = \varepsilon .$$

Thus, the electromotive force of the element equals the work that the unit electric mass would perform during its passage through the electric circuit <if the current strength were very small.

If the current strength is finite, then it always holds that> $P_1 - P_2 + P_4 - P_5$ is a definite quantity independent of i and consequently also of the work of the mass 1 during the passage.

$$(P_1 - P_5) + (P_2 - P_1) + (P_3 - P_2) + (P_4 - P_3) + (P_5 - P_4) = 0$$

$P_1 - P_5 = \text{const.}$ $\qquad$ $P_3 - P_2 = \text{const.}$ $\qquad$ $P_4 - P_3 = \text{const.}$

$$(P_2 - P_1) + (P_5 - P_4) = \text{Const.}$$

But this is the work done by mass 1 [during] one cycle.

Explanation of contact electricity.

We put forth the following hypothesis:

Every particle of matter exerts an attractive force of a definite magnitude on every positive electrical particle which is a function of the distance between the two particles and is proportional to their masses. <The very small sphere of action of this effect of force.>

We consider the effect of these forces on an electrical particle in the interior of a homogeneous mass and on the boundary of two media.

[Fig.]

In case I the particle is in equilibrium and so it is in case 2. However, at a dist[ance] from the wall that is smaller than the radius of the sphere of action, there will result an attraction toward the interior of the matter B if the latter exerts a greater attraction than the matter A. Hence, a flow of positive electricity will occur toward B and a streaming of negative electricity toward A.

Let us consider the force acting at some position on the mass 1 of positive electricity during this process

neg $\qquad$ [Fig.] $\qquad$ pos $\qquad$ $F = -\dfrac{\partial P}{\partial n} + f_1(x) + f_2(x)$

In time, the accumulation of positive masses increases. Thus also P, and thus $\frac{\partial P}{\partial n}$, $f_1(x)$ and $f_2(x)$ represent the acting forces originating from the matter and thus are constant at a fixed position. Hence there must occur a moment at which $F = 0$ or

$$\frac{\partial P}{\partial x} = f_1(x) + f_2(x). \qquad (1)$$

The cumulation then comes to an end, equilibrium sets in.

f_1 and f_2 denote the forces of attraction exerted by the media B and A, resp[ectively]. The course of f_1 and f_2 can be represented graphically in the following manner:

potential curve $\qquad$ [Fig.]

We now multiply equation (1) by dx and integrate between G_2 and G_1, and thus obtain

[180]

$$\int dP = \int f_1(x)\,dx + \int f_2 x\,dx$$

$$P_B - P_a = F_1 - F_2.$$

F are characteristic constants of the materials. Thus, P increases in the boundary layer to a higher value.

Finally we determine the density of the surface layer using Gauss's theorem.

[Fig.] $$\int N\,df = 4\pi\sigma = f_1 0 - f_2(0)$$

$$\sigma = \frac{f_1 0 - f_2 0}{4\pi}$$

Modification of Ohm's law in the case of an electromotive force connected with the circuit.

Using the above definition we have

[Fig.] $$i(w + w') = \mathcal{E}$$

$$i(w_1 + w_2 + w_3 + w') = \mathcal{E}.$$

According to Ohm's law:

$$iw_2 = P' - P''$$

$$i(w_1 + w_3 + w') = \mathcal{E} + (P'' - P')$$

$$iW = \mathcal{E} + P'' - P'.$$

This law is a generalization of Ohm's law. $\mathcal{E}$ can have either sign, depending on whether, if shorted, it would produce a current in the same direction as in the actual case, or in the opposite direction.

Kirchhoff's laws.

1) At any moment the algebraic sum of the currents flowing toward a current branch = 0,

[Fig.] $$(i_1 + i_2 + i_3 + i_4)dt = dP\cdot c.$$

Since c is very small and $\frac{dP}{dt}$ within certain limits, one always has with very good approximation

$$i_1 + i_2 + i_3 + i_4 = 0.$$

2) Within any chosen circuit, every closed polygon obeys the law:

$$\Sigma\, iw = \Sigma\, \mathcal{E}$$

where $\mathcal{E}$'s are to be taken as positive or negative depending on whether they act in the direction of the current or against it.

[Fig.]

By repeated application of Ohm's law

$$\begin{array}{ll} P_1 - P_2 + \mathcal{E}_1 = i_1 w_1 & + \\ P_2 - P_3 = i_2 w_2 & + \\ P_4 - P_3 + \mathcal{E}_2 = i_3 w_3 & - \\ P_1 - P_4 - = i_4 w_4 & - \end{array}$$

$$\mathcal{E}_1 - \mathcal{E}_2 = i_1 w_1 + i_2 w_2 - i_3 w_3 - i_4 w_4$$

$$\Sigma\ \mathcal{E} = \Sigma\ iw.$$

where everywhere one has to take the absolute sum. Kirchhoff 1845.

Application to the determination of current divisions.

[Fig.]

$$i_0 = i_1 + i_2$$

$$\mathcal{E} = i_0 w_0 + i_1 w_1$$

$$0 = i_2 w_2 - i_1 w_1$$

$$i_0 = \mathcal{E}\ \frac{w_1 + w_2}{w_0 w_1 + w_1 w_2 + w_2 w_0}$$

$$i_1 = \mathcal{E}\ \frac{w_2}{N}$$

$$i_2 = \mathcal{E}\ \frac{w_1}{N}$$

$$\Delta P = \mathcal{E}\ \frac{w_1 w_2}{N}.$$

2d case

The junctions yield the following independent equations.

[Fig.]

$$\left.\begin{array}{l} i_0 = i_1 + i_3 \\ i_2 = i_1 + i \\ i_3 = i + i_4. \end{array}\right\}$$

The circuits also yield the eqs.

$$\left.\begin{array}{l} \mathcal{E} = i_0 w_0 + i_1 w_1 + i_2 w_2 \\ 0 = i_3 w_3 + iw - i_1 w_1 \\ 0 = i_4 w_4 - i_2 w_2 - iw \end{array}\right\}$$

$$i_3 = \mathcal{E} \cdot \frac{ww_2 + w_1[w_4 + w_2 + w]}{N}$$

$$i_4 = \mathcal{E} \cdot \frac{ww_1 + w_2[w_3 + w + w_1]}{N}$$

$$i = \mathcal{E} \cdot \frac{w_1 w_4 - w_2 w_3}{N}.$$

i can be adjusted in such a way that i becomes 0. Then we have

$$w_1 w_4 = w_2 w_3$$

$$\text{or} \qquad \frac{w_1}{w_2} = \frac{w_3}{w_4}$$

$$N = w_0 w_3 w_4 + w_0 w_3 (w + w_1) + w_0 w_4 [w + w_1] + w_3 w_4 (w_1 + w_2) +$$

$$(w_0 + w_3 + w_4)(w_1 w_2 + w w_1 + w_0 w_2).$$

Parallel circuits.

By repeated application of Kirchhoff's laws one obtains

$$I_1 = i_1 + I_2 \qquad \text{[Fig.]}$$

$$I_2 = i_2 + I_3$$

$$I_3 = i_3 + I_4$$

$$\varepsilon = I_1 W_1 + i_1 w_1$$

$$0 = I_2 W_2 + i_2 w_2 + I_2 W_7 - i_1 w_1$$

$$0 = I_3 W_3 + i_3 w_3 + I_3 W_6 - i_2 w_2$$

$$0 = I_4 (W_4 + w_4 + W_5) - i_3 w_3.$$

Three-wire system.

[Fig.]

Again one obtains 7 equations and 7 unknowns. *One* of the junctions must always be omitted.

Resistance of the branched conductor.

We denote the quantity $\frac{\varepsilon}{\Sigma i}$ as the latter's resist.,analogously to the [un]branched conductor.

$$W = \frac{<\varepsilon>P_1 - P_2}{\Sigma i} = \frac{\Delta P}{\Sigma \Delta P w} = \frac{1}{\Sigma \frac{1}{w}}.$$

For the case under consideration

$$W = \frac{1}{\frac{1}{w_1} + \frac{1}{w_2}}. \qquad \text{[Fig.]}$$

This leads to a method of dividing a given resistance w into *n* eq. parts.

$W = \frac{w}{n}$. [Fig.]

Measurement of I, w and ε.

Measurement of W.

One uses the Wheatstone bridge.

[Fig.]

<According to> Let AC be the resistance w that has to be measured. CB the known resistance of magnitude n. According to the above,

$$i = \varepsilon \frac{(W_1W_4 - W_2W_3)}{N}.$$

If $i = 0$, then we must also have $W_1W_4 - W_2W_3 = 0$,

$$\text{hence} \quad \frac{W_1}{W_2} = \frac{W_3}{W_4} = \frac{w}{n}$$

$$w = n\frac{W_1}{W_2}.$$

Hence if one knows $\frac{w_1}{w_2}$ in this case, then w is known as well.

Based on Kirchhoff's idea, one uses a homogeneous wire of a constant cross section as the conductor AB. Then we have

$$w_1 = \frac{\omega}{q} \cdot \ell_1$$

$$\text{hence } \frac{w_1}{w_2} = \frac{\ell_1}{\ell_2}.$$

$$w_2 = \frac{\omega}{q} \cdot \ell_2$$

Hence, if one determines the location D by appropriate shifting of a slide, and measures ℓ_1 and ℓ_2, one obtains

$$w = n \cdot \frac{\ell_1}{\ell_2}$$

The absence of current in the conductor CD is demonstrated using an inserted galvanometer.

Determination of D by interpolation

[Fig.]

We investigate how the current in CD [varies] with the distance $DD' = x$ from the zero-current point.

$$0 = \mathcal{E}\,\frac{(w_1 w_4 - w_2 w_3)}{N}$$

$$i_X = \mathcal{E}\,\frac{\left[w_1\left(w_4 + \frac{x}{q}\,\ell\right)\right] - \left[w_2\left(w_3 - \frac{x}{q}\,\ell\right)\right]}{N_X}$$

$$= \frac{(w_1 w_4 - w_2 w_3) + \alpha x \ldots}{A + Bx \ldots}\,.$$

Neglecting the terms of higher than 1st order in x,

$$i_X = \alpha x(A - B)x = cx.$$

i_X is measured by the tangent galvanometer. If its deflection is v, then $i_X = c'\ \mathrm{tg}\ v$.

From this it follows that $i_X = Cv$, substituting the arc for the tangent. If i_X is measured for 2 positions x_1 and x_2, one obtains the true neutral position, i.e., the p[oint] $x = 0$, on account of proportionality, by using the following construction.

[Fig.]

This measurement contains a source of error due to the circumstance that the resistance of the endpoints of the measuring wire cannot be taken into account, i.e., the beginning of the measuring scale does not correspond to the beginning of the conductor. This can be avoided by the following arrangement.

[Fig.]

With this arrangement we make the three indicated connections between the little dishes and the measuring wire in succession and determine for each the point A on the measuring wire at which the current is zero. According to the foregoing we then have the following relation.

$$\frac{\alpha}{W + N + \beta} = \frac{x + W(AA_1)}{W(A_1\mathcal{E}) + z} \qquad \Bigg| \qquad \frac{W + N + \beta}{W + N + \alpha + \beta} = \frac{W(A_1\mathcal{E}) + z}{x + z + W(A\mathcal{E})}$$

N_1 analogously N_2

$$\frac{\alpha + W}{N + \beta} = \frac{x + W(AA_2)}{W(A_2\mathcal{E}) + z} \qquad \text{or} \qquad \frac{N + \beta}{N_1} = \frac{W(A_2\mathcal{E}) + z}{N_2}$$

$$\frac{\alpha + W + N}{\beta} = \frac{x + W(A_1A_3)}{W(A_3\mathcal{E}) + z} \qquad \Bigg| \qquad \frac{\beta}{N_1} = \frac{W(A_3\mathcal{E}) + z}{N_2}.$$

$$\frac{W}{N_1} = \frac{W(A_1A_2)}{N_2} \qquad \text{or} \qquad \frac{W}{N} = \frac{W(A_1A_2)}{W(A_2A_3)}$$

$$\frac{N}{N_1} = \frac{W(A_2A_3)}{N_2}$$

Thus, if the measuring wire has a constant cross section and is homogeneous, we also have

$$\frac{W}{N} = \frac{\ell_1}{\ell_2}.$$

Thus, three [different] readings eliminate the source of error.

During the measurement the current must be of short duration, otherwise the heating due to the current flow would considerably increase the resistance.

The relation between specific resistance and temperature.

One must here distinguish between a dependence on the physical and on the chemical structure of the metal. The former play[s] a secondary role in comparison with the latter.

Pure metals obey a simple law. If one makes a series of resistance measurements at various temperatures for a wire of a completely pure substance, one obtains a strictly linear relation between W and t

$$W_t = W_0(1 + \alpha t)$$

$$\alpha = \frac{W_1 - W_0}{W_0 t}.$$

Thus, it suffices to conduct the measurement at temperatures 0 and t.

Different metals yield

α between 0.0038 and 0.0042.

This value roughly agrees with the coefficient found in the investigation of the relationship betw[een] temperature and pressure of an ideal gas. $(\frac{1}{273})$ $p = p_0(1 + \alpha t)$. Thus, it turns out that the conductivity of pure metals would become ∞ large at the same temperature at which the gas pressure becomes 0,, provided that the linear relationship still holds for low temperatures. The assumption that this is indeed so was confirmed in an investigation of the resistance of metals at the temperature of liquid atmospheric air (-192°). Thus, it is almost certain that the posited 0-point of temperature corresponds to reality.

Alloys behave completely differently. Even though the absolute value of their specific resistance is considerably higher than that of pure metals, it varies much less with the temperature, often only 1%.

Constantan	60% Cu & 40% Ni	$\alpha = 0.0001$
Manganin	84 Mg 12 Cu 4 Ni	$\alpha = 0.000011$

Relatively minute admixtures of metals or metal oxides suffice to increase the specific resistance of a metal significantly.

The following quantities are required for the numerical measurement of the resistance

ω, ℓ, q, t, where ℓ meas[ured] in m
q in mm^2
and t in deg. Celsius.

Thus, it remains to establish the unit of ω.

After the most different kinds of units had been used for a long time, Siemens proposed the following as the unit:

$\omega = 1$ is the resistance of a mercury column of 1 mm^2 cross section and 1m length at 0° Celsius.

Later on one undertook to fit this unit into the c.g.s. system (Paris Congress. This unit was established based on the caloric measurements of Peltier's effect. It turned out that Siemens' unit had the value of $0.95 \cdot 10^9$ in the old system of units. Therefore, instead of Siemens' mercury column, an analogous column with a length of 1.06 m (later on corrected to 1.063) was established as the practical unit, which was named 1 Ohm (Ω).

Production of the Ohm.

Let us have a mercury column in a glass tube of length L (in cm) and cross section q (in mm). We prepare a metal wire with a resistance of 1 Ohm.

For that purpose one chooses a metal whose specific resistance is as constant as possible, say one of the familiar alloys.

The resistance of the glass tube is $W = \frac{L \text{ (in cm)}}{q \text{ in mm } 106.3}$

We set up the familiar arrangement.

[Fig]

The resistance x shall become equal to 1.

$$\frac{W}{x} = \frac{\ell_1}{\ell_2} \text{ when zero current in the bridge}$$

$$x = W\frac{\ell_2}{\ell_1} = \frac{\ell_2}{\ell_1}\frac{L}{q(106.3)}.$$

Now one keeps varying x until the expression on the right hand side = 1 at zero current.

The resistance is then $x = 1^\Omega$.

We have then $\frac{\ell_1}{\ell_2} = \frac{L}{q\ (106.3)}$.

If one has prepared a number of such wires of resistance 1, it is easy also to prepare resistances of other mag[nitudes] and $\overset{2}{\frac{1}{2}} \quad \overset{3}{\frac{1}{3}} \quad \overset{4}{\frac{1}{4}}$

by connecting them in series and in parallel.
Siemens devised a convenient assembly of resistance units according to the following scheme.

[Fig.]

Using metal stoppers it is possible to produce resistances of magnitude 1 - 10.

To produce resistance wires, one again uses alloys. The measured value has to be corrected for temperature. If the res. is calibrated at 0°, then $W = W'(1 + \alpha t)$, where W' is the resistance read off, and W the actual resistance. To avoid induction effects and errors of measurement associated with them as much as possible, the wire is bent in the indicated manner before winding and is taken double.

[Fig.] $\qquad W^{\Omega} = \omega \cdot \frac{\ell}{q}.$

In practice, ℓ is us[ually] measured in meters and q in mm².

The sp[ecific] resistance depends strongly on the chemical purity of the substance.

E.g.

$Cu + \frac{1}{1000}$ Zn....<10>..5% increase of ω

$Cu + \frac{1}{1000}$ Pb... 11% " " "

$Cu + \frac{1}{10000}$ P 60-70% " " "

An example of the difference in resistances between pure metals & alloys.

1 part platinum plus 2 parts silver produce an alloy of

$$\omega = 0.24$$

$$\omega_{\text{platinum}} = 0.059$$

$$\omega_{Ag} = 0.0187.$$

The connection between electrical and thermal conductivities of pure metals.

If one measures thermal conductivity using the units

gr
cm
minutes
1°

one obtains almost the same numbers as when one measures q using the units given above ($q = \frac{1}{\omega}$).

The thickness of wires is measured using micrometers.

It is possible to read off $\frac{1}{2000}$ mm.

Measuring the temperature by measuring the resistance.

Since measurement of resistance is to be done with precision, and resistance is related to temperature in a simple way and varies with temperature relatively quickly, it can be used for the determination of the temperature.

To measure the temperature at locations that are not easily accessible because of their distance or other circumstances, <the following> one uses a method of measurement according to the following scheme:

[Fig.]

We vary the resistor N until $\ell_1 = \ell_2 = \ell$. Further, let us have

$$K_1 = K_2 = K_3 = K.$$

Then:

$$\frac{N + K}{w + K + n} = 1.$$

$N = w + n$, where n is known.
" N " "

Thus we obtain w. However,

$$w = w_0(1 + \alpha t)$$

$$t = \frac{w - w_0}{w_0 \alpha}.$$

The temperature at the bottom of the sea, for example, can be measured in this way, as well as the temperature in blast furnaces or chimneys.

The measuring wire w is then to be chosen in the following way.

[Fig.]

porcelain

The electrical method of measurement enables us to measure minute temperature differences and is therefore suitable for the detection and measurement of heat radiation.

How [A?] sensitive resistance w. is then built by using a large number of finely rolled-out metal strips (tin foils) connected in series.

[Fig.]

We investigate the course of the process with one-sided irradiation.

Prior to having the radiation act, we adjust the bridge wire to zero current.

We have then

$$W_1 (W_4)_0 - W_2 W_3 = 0.$$

If one lets the radiation act, W_4 changes and one obtains a current I in the bridge.

We have then

$$I = \varepsilon \frac{W_1 W_4 - W_2 W_3}{N}.$$

However, $W_4 = (W_4)_0 (1 + \alpha t)$.

If this is substituted, the numerator and denominator will change.

$$I = \varepsilon \frac{(W_1{}^2 W_{40} - W_2 W_3) + W_1 (W_4)_0 \alpha t}{N + At} = \varepsilon \frac{W_1 W_{40} \alpha t}{N} \left[1 - \frac{At}{N}\right].$$

Since the temperature difference is supposed to be small, <apparently> one can put $1 - \frac{\alpha t}{N} = 1$, thus very closely

$$I = \varepsilon \frac{W_1 W_{40}}{N} \alpha t .$$

We have, however,

$$dW = SO' dz - hO_{tot} \cdot t dz = Mc\ dt .$$

O' is the irradiated surface, S the radiation intensity, i.e., the quantity of heat striking in unit time the unit surface area whose normal has a direction opposite to that of the rays, t is the temperature difference relative to the surroundings.

$$dt = \left[\underbrace{\frac{SO'}{Mc}}_{a} - \underbrace{\frac{hO_{tot}}{Mc}}_{b} t\right] dz$$

$$dz = \frac{dt}{a - bt}$$

$$z + C = - \frac{1}{b} \lg (a - bt).$$

The boundary cond[ition] yields

$$C = - \frac{1}{b} \lg a$$

$$\overline{z = \frac{1}{b} \lg \left[\frac{a}{a - bt}\right]}$$

$$\frac{a}{a - bt} = e^{bz}$$

[198]

$$\frac{a}{b}(1 - e^{-bz}) = t$$

$$t = \frac{SO'}{hO_{tot}}\left[1 - e^{-\frac{hO}{Mc}z}\right].$$

Since M and c are very small quantities, the temperature very quickly reaches the stationary value

$$t = \frac{SO'}{hO_{tot}}.$$

If HO and O' are known, a precise measurement of S can be carried out in this way in an absolute manner, otherwise, by comparing two stationary temperatures at different irradiations, a relative one.

Thus, there exists a direct relation between the deflection of the galvanometer x and the radiation S.

$$Cx = I$$

$$I = \varepsilon \frac{W_1 W_{4_0}}{N} \alpha t$$

$$t = \frac{SO'}{HO}$$

thus $$<S>Cx = \frac{\varepsilon}{N} W_1 W_{4_0} \alpha \frac{O'}{<H>hO} S.$$

By sensitivity <one understands> of the galvanometer one understands the current in amp. necessary to produce a deflection of 1 mm. The sensitivity of the galvanometer in the Z[urich] laboratory is $\frac{1}{10^7} - \frac{1}{10^{10}}$. With an appropriate choice of W_1 and W_4 and ε it is possible to detect temperature differences of $\frac{1}{10^5} - \frac{1^\circ}{10^6}$.

Resistance of copper.

$$\omega = 0.0157$$

Here the units are

$$\frac{\mathrm{m}}{\frac{\mathrm{mm}^2}{\Omega}}$$

If cm is introduced as the unit in length and cross section, we obtain $\omega = 0.00000157$.

Hence in absolute units

$$\omega = 1570.$$

This number applies for 0°.
For the temperature t we have $\omega = \omega_0(1 + 0.0041t)$.

$$\boxed{\omega_{15} = \frac{1}{<76>60}} \qquad \text{(approximately).}$$

Measurement of the current.

A direct measurement of the current, i.e., of the quantity of electricity passing through the cross section of the conductor in unit time, is not feasible. The current must therefore be determined from its effects. For that purpose, one uses almost exclusively the magnetic effect.

Again we start from a hypothetical fundamental law and compare the consequences with experience.

If a magnetic mass is brought near a part of an electric circuit, a force will be exerted upon it perpendicular to the direction of the current element & ⊥ to the direction of the connection between the mass and the current element.

M_{north} [Fig.]

The direction of this force is obtained in the following way. An observer swimming with the current with his head forward sees the force exerted on the north magnetic mass as directed to the left and that exerted on the south magnetic mass in the opposite direction.

In accordance with the principle of the equality of action and reaction, the magnetic mass exerts a force of the opposite direction on the current element .

This is demonstrated by the arrangement (Ampère) shown in the accompanying scheme.

[Fig.]

The elementary law of the electromotor magnetic force was established by Laplace on the basis of Biot's and Savart's investigation on the magnetic effect of long linear currents.

It reads

[Fig.]

$$df = \frac{i\ ds\ m}{r^2} \sin \varphi .$$

The direction is as indicated above.

A direct proof for this law is not feasible but the conclusions deduced from it are in complete agreement with experience.

The expression of the law in the Cartesian coordinate system.

$$dX = \frac{i\ ds\ m}{r^2} \sin \varphi\ \alpha.$$

[Fig.]

$$dY = \frac{i\ ds\ m}{r^2} \sin \varphi\ \beta.$$

$$dZ = \frac{i\ ds\ m}{r^2} \sin \varphi\ \gamma.$$

The direction of the force is ⊥ to r and to ds, hence

$$a\alpha + b\beta + c\gamma = 0$$
$$a'\alpha + b'\beta + c'\gamma = 0$$

furthermore

$$\alpha^2 + \beta^2 + \gamma^2 = 1$$

hence,

$$\alpha = \frac{bc' - cb'}{\sqrt{(bc' - cb')^2 + (ca' - ac')^2 + (ab' - ba')^2}}$$

$$\beta = \frac{ca' - ac'}{\sqrt{\qquad}} \qquad\qquad \gamma = \frac{ab' - ba'}{\sqrt{\qquad}}$$

$$\cos \varphi = aa' + bb' + cc'$$

$$\sin \varphi = \sqrt{(a^2 + b^2 + c^2)(a'^2 + b'^2 + c'^2) - (aa' + bb' + cc')^2}$$

$$= \sqrt{(ab' - ba')^2 + (bc' - cb')^2 + (ca' - ac')^2}$$

Hence

$$dX = <\mp> - \frac{i\ ds\ m}{r^2} (bc' - cb')$$

$$dY = <\mp> - \frac{i\ ds\ m}{r^2} (ca' - ac')$$

$$dZ = <\mp> - \frac{i\ ds\ m}{r^2} (ab' - ba')$$

[Fig.]

The diagram shows that the sign we have chosen is correct for a coordinate system of the kind indicated. To simplify the formulas, we choose, in analogy with the treatment of electrical masses, the south magnetic masses with negative signs.

The galvanometer.

Let a steel ring of the indicated form be exposed to a magnetic field of the indicated direction.

[Fig.]

It then will become magnetized, i.e., the magnetic masses in it will become permanently separated. It is then possible (which cannot be proved here) to specify two magnetic masses of different sign in diametrically opposite points of the center, at a distance λ from it, which can completely replace the ring magnet as far as magnetism is concerned. These points are called the poles of the magnet.

Let this ring be suspended for torsion so that the connection $m - m$ always remains horizontal.

Let the ring be surrounded by a circular current of radius R, whose dimensions should be very large compared with those of the ring magnet. The plane of the circuit shall be vertical.

φ runs from 0 to 2π

[Fig.]

$$ds: \quad x = 0, \quad y = R \sin \varphi, \quad z = R \cos \varphi$$

$$m: \quad x = \lambda \sin u, \quad y = \lambda \cos u, \quad z = 0$$

$$r = \sqrt{\lambda^2 \sin^2 u + (R \sin \varphi - \lambda \cos u)^2 + R^2 \cos^2 \varphi}$$

$$= \sqrt{R^2 + \lambda^2 - 2R\lambda \sin \varphi \cos u}$$

$$= R \sqrt{1 - 2\frac{\lambda}{R} \sin \varphi \cos u} \quad \text{(neglecting the 2nd power of } \tfrac{\lambda}{R}\text{)}$$

$$= R \left(1 - \frac{\lambda}{R} \sin \varphi \cos u\right).$$

The direction cosines

	a	b	c
(r):	$\dfrac{-\lambda \sin u}{r}$	$\dfrac{R \sin \varphi - \lambda \cos u}{r}$	$\dfrac{R \cos \varphi}{r}$
	a'	b'	c'
(ds):	0	$\cos \varphi$	$- \sin \varphi$

$$dX = \frac{mi\ ds}{r^2} \frac{R}{r} \left\{ \left(\sin \varphi - \frac{\lambda}{R} \cos u\right) \sin \varphi + \cos^2 \varphi \right\}$$

$$<=> \frac{im\,ds}{r^2} = \frac{imR\,d\varphi}{r^2}.$$

dY and dZ become <infin> small of higher order of $\frac{\lambda}{R}$ and are to be neglected.

$$r^2 = R^2 \left[1 - 2\frac{\lambda}{R} \sin \varphi \cos u\right]$$

$$dX = \frac{im\,d\varphi}{R} \left[1 + 2\frac{\lambda}{R} \sin \varphi \cos u\right]$$

$$X = \int_0^{2\pi} \frac{im}{R} \left[1 + 2\frac{\lambda}{R} \sin \varphi \cos u\right] d\varphi .$$

The second term vanishes, so that we have

$$X = \frac{2\pi i m}{R}.$$

Now we formulate the torque of the electromagnetic force.
The torque of both forces is

$$2\lambda X \cos u = \frac{4\pi \lambda i m}{R} \cos u .$$

[Fig.]

The quantity $m \cdot 2\lambda$ is called the magnetic moment M. Then we have

$$\text{Torque} = M \cdot \frac{i}{R} \cdot 2\pi \cos u .$$

In addition, however, the magnet is also acted upon by the terrestrial magnetic force. As the field of force of strength 1 we define the field which exerts force 1 in absolute units on the unit magnetic mass. The horizontal component of the terrestrial magnetic force, which is the only one to exert an effect in our case, we denote by H, and imagine that it has been measured absolutely by a procedure which will be described later. We further imagine that the apparatus is oriented such that the plane current circuit coincides with the magnetic meridian.

The force acting upon the north end is mH.

[Fig.]

Hence its torque is $mH\lambda \sin u$.

Thus, the total torque arising from the terrestrial magnetic force, which acts in the same or the other direction than the electromagnetic force of the current circuit, is

$$\text{Torque} = 2mH\lambda \sin u = <2> M<\lambda>H \sin u .$$

These two torques must be equal in equilibrium. Hence,

$$\not{M}H \sin u = \not{M}i \frac{2\pi}{R} \cos u$$

or $$i = H \frac{R}{2\pi} \cdot \mathrm{tg}\ (u).$$

If H is known in absolute units, then thereby the absolute unit of i is also defined.

One can posit: $i = 1$ when the right-hand side of the expression = 1.

The Paris Congress established an additional, so-called practical unit, which is 10 times smaller.

The relation then becomes

$$i = 10H \frac{R}{2\pi} \mathrm{tg}\ u.$$

$\frac{R}{2\pi}$ is called the constant of the galvanometer. In our region H equals about 0.21 in |g cm sec|. The absolute unit of current has been introduced by Wilhelm Weber.

The reading method of Gauss and Poggendorf.

One gets

$$\mathrm{tg}\ (2u) = \frac{a}{D}.$$

[Fig.]

Were one e.g., to choose $D = 2m$, then a rotation of 1° would yield

$$\mathrm{tg}\ 2u = 0.0348$$
$$\mathrm{tg}\ 2u \cdot D = 70 \text{ mm}.$$

If readings can be made with an accuracy of $\frac{1}{10}$ mm, then it follows that an accuracy of $\frac{1}{700}$° is attainable for this distance.

Intervals in which currents can be measured in this way are relatively few.

To be able to measure small currents, one only needs to multiply their number of windings. Each winding then exerts an almost equal effect on the ring magnet so that one obtains

$$i = H \frac{\bar{R}}{2\pi n} \mathrm{tg}\ u$$

where $\bar{R}$ denotes the average radius of the windings.

Large currents.

Only small currents can be measured directly (without current branching) with the apparatus described because one cannot choose a value of tg $(2u)$ exceeding $\frac{1}{4}$. For that purpose we use eccentric current circuits.

[Fig.]

By a calculation analogous to that above, one obtains

$$\text{Torque} = Mi \frac{2\pi R^2}{\sqrt{R^2 + a^2}} \cos u$$

The counterac[ting] torque is again

$$M \cdot H \sin u$$

$$\langle I \rangle i = \left\langle H \frac{R^2}{(\sqrt{R^2 + a^2})} \right\rangle H \cdot \frac{(\sqrt{R^2 + a^2})^3}{2\pi R^2} \operatorname{tg}(2u).$$

Considerable current strengths can be measured in this way, especially if one has two movable circuits.

One has t[hen]

$$\text{Torque } I - \text{Torque } II = HM \sin u.$$

One gets

$$i = \frac{H}{2\pi R^2} \frac{1}{\frac{1}{(\sqrt{R^2 + a_1^2})^3} - \frac{1}{(\sqrt{R^2 + a_2^2})^3}} \operatorname{tg} u.$$

Large currents can be measured directly in this manner.

[Fig.]

38. TO MAJA EINSTEIN

[Zurich, 1898]

If things had gone my way, Papa would have looked for employment already 2 years ago, and he and we would have been spared the worst... What depresses me most is, of course, the misfortune of my poor parents who have not had a happy moment for so many years. What further hurts me deeply is that as an adult man, I have to look on without being able to do anything. After all, I am nothing but a burden to my family... It would indeed be better if I were not alive at all. Only the thought that I have always done whatever lay within my modest powers, and that year in, year out I do not permit myself a single pleasure, a distraction save that which my studies offer me, sustains me and must sometimes protect me from despair.

39. TO MILEVA MARIĆ

Zurich Wednesday [16 February 1898]

Esteemed Miss!

The desire to write to you has finally overcome the bad conscience that plagued me because I haven't replied to you for such a long time & that made me avoid exposing myself to your critical eyes. But even though you are rightfully a little angry with me, there is one thing you have to give me credit for, namely that rather than adding to my sin by taking shelter behind poor excuses, I am asking you simply and straightforwardly to forgive me - and answer me as soon as possible.

I am very happy about your intention to continue your studies here again, just do it pretty soon; I am sure you will not regret it. I am quite convinced that you will be able to catch up in a relatively short time with the main courses we had. To be sure, it puts me in a very embarrassing position if I have to tell you what material we covered. Simply, it's only here that you will find the material properly arranged and elucidated.

Hurwitz lectured on differential equations (exclusive of partial ones), also on Fourier series, a little on calculus of variations & double integrals. Herzog [on] dynamics and strength of materials, on the latter very lucidly and well -- on dynamics somewhat superficially, which is quite natural with a "mass course." Weber lectured on heat (temperature, heat quantities, thermal motion, dynamic theory of gases) with great mastery. I am looking forward from one of his lectures to the next. Fiedler lectures on projective geometry, he is the same indelicate, rude man as before & in addition sometimes opaque, but always witty & profound -- in brief, a master but, unfortunately, also a terrible schoolmaster. The only other course of importance that will give you a lot to do is the theory of numbers, but you can make up this subject by studying privately at your leisure.

If I am allowed to give you some advice (completely unselfishly?), it's that you come here as soon as possible, because here you will find everything you need compressed in our notebooks.

As a precaution, you could write to Hurwitz in advance and ask him whether he agrees. I believe that, sooner or later, it will be possible for you to lodge again with the Bächtholds, because one room has not been definitively rented. Of course, that certain Zurich philistine is now living in your former pleasant little room & you must renounce it.... serves you right, you little runaway!

But now I must push on with my studies

With cordial greetings your

Albert Einstein

40. TO MILEVA MARIĆ

[Zurich, 16 April - 8 November 1898]

D[ear] M[iss] M[arić]!

Please don't be angry that I didn't show up for such a long time. I was seriously unwell, so that I didn't dare leave my room. My legs are still somewhat unsteady today. Nevertheless, I was bold enough to pull myself together for a walk this afternoon. I told Mrs. Bächth[old] to invite all those boarders who feel like it & I beg of you to be one of them. But if you don't come, I'll visit you very soon, once I feel well enough. If I am not able to come, then I expect that you'll visit me as soon as possible.

With friendly greetings, your

Albert Einstein

I've already finished reading half of the volume. I find it very stimulating & informative, even though the details sometimes lack clarity & precision.

41. TO MILEVA MARIĆ

[Zurich, after 16 April 1898]

Esteemed Miss!

When I got home just a short while ago, I found the apartment locked & nobody home, so that I again had to depart in humiliation.

I beg you therefore not to be angry with me for abducting Drude in my hour of need so as to be able to study a little. Friendly greetings from your

Albert Einstein

42. JÉRÔME FRANEL TO HERMANN BLEULER

Zurich, 21 October 1898

Mr. President of the Council of the Polytechnikum in Zurich

Mr. President,

The enclosed table contains the marks obtained by the students of the sixth department, section A, who have taken the intermediate diploma examinations (Uebergangsdiplom). Naturally, in view of the excellent results, the members of the conference have the honor to propose to the School Council that all candidates be permitted to take the final diploma examination.

Please accept, Mr. President, my expressions of highest regard.

J. Franel
Deputy Principal,
the sixth department, section A

Intermediate Diploma Examinations
Sixth Department, Section A

	Differential & Integral Calculus	Analyt. Geometry	Descrip.Geom. & Projective Geometry	Mechanics	Physics	Avg. Grade
Ehrat	*5*	*5*	*5½*	*5*	*5½*	*5.2*
Einstein	*5½*	*6*	*5½*	*6*	*5½*	*5.7*
Grossmann	*5½*	*5½*	*6*	*5½*	*5½*	*5.6*
Du Pasquier	*5½*	*5*	*5*	*5*	*6*	*5.3*
Kollros	*5½*	*5½*	*5½*	*5½*	*6*	*5.6*

approved
22 Oct 98
H Bleuler

43. TO MILEVA MARIĆ

[Zurich, after 28 November 1898]

Esteemed Miss!

Marco Besso died Sunday night. A terrible blow to his family, but still better than a miserable life.

If it is alright with you, I'll come to your place this evening to read. Your

Albert Einstein

44. TO MAJA EINSTEIN

[Zurich, after February 1899]

There is enough to do, yet not excessively, so that sometimes I find the time to idle away an hour or so in the beautiful surroundings of Zurich. In addition I am happy at the thought that the worst worries are over for my parents. If all people would live in such a way (namely like myself), writing of novels would never have been born.

45. TO MILEVA MARIĆ

[Milan] Monday [13 or 20 March 1899]

Dear Said!

A good and proper scolding I just received brought you vividly to my mind, which will now immediately be documented by a little letter.

To the paragon, a sample (of course of no value) [untranslatable wordplay referring to a "sample without value" parcel]. Has it arrived by now. If it has not, you mustn't beam in anticipation. It is not for eating. Oh, yes -- in emulation of famous examples, the letter for the paragon is also put into the sample [a triple pun on "Muster"] -- dared is half won.

I am having a terrific time at home; I have been passing much of the time in inmost delights, that is to say, I have been eating quite a bit & quite well, so that I already suffered a little from our favorite poetic sorrows, like then at Sterns', when I was sitting for hours next to my fascinating charming [female] dinner partner. It was then that it struck me most vividly how closely our mental and physiological lives are linked.

The journey was very nice, even though, unfortunately, all the fellow-passengers were male creatures. There were, for example, a few nice and lively Italian youngsters, who sang & laughed & joked among themselves, half like young girls, half like puppies. I did fine in Chiasso. This fellow has nothing ???, the shrewd [customs] officer must have thought. While I most earnestly conversed about Italian conditions with a young man during the continuation of the trip, a German youth & business apprentice, who was traveling to Italy for the first time, was taking pains to dispose of the few occasional scraps of Italian, which he had acquired especially for this purpose, as elegantly and casually as possible. This was as if someone with a trumpet that has only 2 notes wants to play in an orchestra & is ardently waiting for the moment when he can again sound off one of them. Your photograph made a great impression on my old lady. While she was immersed in its contemplation, I added with deepest understanding: Oh, yes, she is a clever creature. I had already to endure quite a lot of teasing for this & for similar things, but I don't find this the least bit disagreeable.

My broodings about radiation are starting to get on somewhat firmer ground & and I am curious myself whether something will come out of them.

Friendly greetings *etc.*, especially the latter, from your

Albert

Regards from my old lady.

46. TO ROSA WINTELER

Zurich <Friday> Saturday [29 April 1899]

Dear Miss!

Before I set about my day's work today I want to thank you for your friendly little card, after I had been waiting to see whether I couldn't find time to visit you. I wanted to go just today -- but I did not finish my work; however, I'll be sure to come a week from Sunday. (I think if I promise it, then it will go easier.)

I wonder how the little Winteler home feels under your management? It probably thinks: Yes, indeed, Rosa is the only one who remains reliable and faithful to me! But all the same, in this it is a little bit mistaken. Little mother-in-law, for example, is not an activity to be sneezed at either: -- And your foster-children, Pauli cat dog little bird, how good will they have it under such strict discipline and attention....And just across the street you have the prime fount of Aarau wisdom, which is tapped fresh from the barrel every day. Do you go there already? If so, then give my regards to Mr. Schuft [nickname meaning "rogue"] & also to Mr. Rennhart -- if the latter still remembers me. As long as we are at it, give also my kindest regards to your friend the little Miss Bride. Ask her also whether her sweetheart is sending her many sentimental little letters -- oh, that must be nice -- if you haven't asked her already (?)

I am just fine. I work hard & feel very pleased & good in doing so. The days & weeks slip by just so, without my noticing it except toward the end of the month.....the old song with the old motif & and the many variations, a few of which you too probably know how to sing, Miss Housewife!

But now I have to start with my work! Kindest regards from your

Albert

47. TO ROSA WINTELER

Zurich, Thursday [18 May 1899]

D[ear] Miss Rosa!

I received a postcard from Mommy announcing that she and Papa will arrive here Friday evening & will immediately continue for Aarau, and that I should inform you of that. If it's at all feasible, I'll come to you at Whitsuntide. I am immensely delighted that once again I'll be able to pleasantly idle away such a day, because I am frightfully overburdened with work these days.

Cordial greetings to everybody there, and especially to you from your

Albert

48. TO JULIA NIGGLI

Zurich, Friday [28 July 1899]

Dear Miss!

When your nice little card was lying on my table the other day, I felt very happy, and I would have felt much ashamed for not yet having responded to your kind invitation had I been able to write. It so happened that I injured my right hand rather seriously in the physics laboratory two weeks ago, so that the wound had to be closed by stitches at the clinic. Writing still causes me considerable trouble. But soon it will be fine again. I have not been to Aarau since, and neither have I been able to write there. In case you correspond with the Wintelers, please don't write them anything about that.

You certainly seem to be in an enchanting little nest at present. I would love to help you kill time there in all sorts of pleasant ways. Unfortunately, my time slips by in everyday philistine work; but my hour of freedom is to strike soon -- and I am immensely happy about that. Starting with the first of August I'll be with my old lady and my sister in Mettmenstetten (line Zurich-Affoltern-Zug); you promised me to visit us there. A man's word is his bond -- which means that you must definitely come! In a week's time I'll be hiking on the Säntis, this will set off some marvelous clambering.

Of course, my fiddle had to be laid aside. I am sure it wonders why it is never taken out of the black case, it probably thinks it has gotten a stepfather. How I miss the old friend through whom I say and sing to myself all that which I often do not at all admit to myself in barren thoughts but which, at best, makes me laugh when I see it in others...

To my greatest regret, I will not be able to go and see the Winteler family for a long time, perhaps in September, if I don't go home to Milan, which I would have first to make plausible to my dear mama. Also, I may be doing scientific work with an Aarau gentleman during the second half of the vacation, and then many a musical hour will probably crop up for us.

With the immodest plea that you write me a few lines when it happens to rain in Strada and you have nothing to do, and with affectionate greetings, I wish you a very enjoyable and good vacation. Your

Albert E.

49. VERSE IN THE ALBUM OF ANNA SCHMID

August 1899

You girl small and fine
What should I inscribe for you here?
I could think of many a thing
Including also a kiss
On the tiny little mouth.

If you're angry about it
Do not start to cry
The best punishment is--
To give me one too.

This little greeting is
In remembrance of your rascally little friend.

Albert Einstein

50. TO MILEVA MARIĆ

<Zurich> Paradies [Mettmenstetten, early August 1899]

D[ear] D[oxerl]

I bet you are surprised to see my hieroglyphs so soon again, knowing what a lazy letter writer I am.

Here in Paradies I live a very quiet, nice, philistine life with my mother-hen & sister, exactly the way the pious & the righteous of this world imagine Paradise to be. Meanwhile, I have already studied quite a bit of Helmholtz on atmospheric movements -- out of fear of you & also for my own pleasure, let me immediately add that I will reread the whole stuff with you. I admire the original, free mind of Helmh. more and more. You, poor soul, must now stuff your head full of gray theory. But I know well both you and your divine composure & know that you'll accomplish all that with a calm mind. Besides, you are at home and are being thoroughly pampered as is proper for a little daughter. To be sure, in our place in Zurich you are the mistress of the house, which is not bad either, & of what a marvelous household at that! When I was reading Helmholtz for the first time, it seemed inconceivable that you were not with me & now it's not much better. I find the collaboration very good & curative & also less desiccating.

I find my mother & sister a bit narrow-minded & philistine, despite all the sympathy I feel for them. It is strange how a way of life will gradually change us and all the nuances of our soul, so that the closest natural bonds of kinship are reduced to a friendship born of habit & that deep inside we become so incomprehensible to each other that we are unable to actively empathize with the other and feel what moves the other.

Now you have just enough to decipher, considering how little time you can now devote to me. If you have time, write me again, but if not, I'll know the reason.

Kindest regards to your d[ear] family, and especially to you from your

Albert

51. TO JULIA NIGGLI

Paradies [Mettmenstetten] Sunday [6? August 1899]

Dear Miss!

Thank you for your friendly letter, which made me very happy, the more so since it gave me for the first time here the happy feeling one experiences when opening a personal letter.

What a strange thing must be a girl's soul! Do you really believe that you could find permanent happiness through others, even if this be the one and only beloved man? I know this sort of animal personally, from my own experience as I am one of them myself. Not too much should be expected from them, this I know quite exactly. Today we are sullen, tomorrow high-spirited, after tomorrow cold, then again irritated and half-sick of life -- and so it goes -- but I have almost forgotten the unfaithfulness & ingratitude & selfishness, things in which almost all of us do significantly better than the good girls...

In fact, I should have spoken only for myself were it not for the sad comfort that most of the others are basically not any different.

Thank you for asking so kindly about my hand, which is now completely cured. I only wanted that you do not mention anything about it to the Wintelers so that they would not worry about me, otherwise I do not want to keep anything secret from them. You may tell them all about it afterwards and also give them my best regards.

It was wonderful on the Säntis. I was up there twice in 4 days. The hike is extremely rewarding and nothing less than difficult. There is nothing nicer for such a stay-at-home as myself than to roam in nature for a change, and open himself to whatever happens to be offered to his eyes instead of always deciding himself what should occupy the mind. I now feel like a harassed housewife on vacation who normally has to prepare the daily menu herself but is now very amiably fed without end in a hotel.

I share your pain that your life at home, with all its prose, must feel even more bitter after the beautiful vacation. At all events, I congratulate you on your decision to go abroad. Every girl should do that if she has the strength to do it. I know already now about *one* job in Italy, which has some good points as far as you are concerned, and, of course, also some bad points, as everything in the world. So, out with it, I will describe to you everything as objectively as I can. You know my cousin Robert Koch. His mother needs a governess for her only daughter (7 years old). My aunt is a woman of natural intelligence, truthful, superficially educated, just, vain, domineering, but also dispassionate and communicative. She expects that people behave properly toward her, but her behavior is also thoroughly correct. But she is rather tactless and insensitive. She is the mistress and the master of the house. The food is good -- she is a capable housewife. You would thus have to take care *only* of the child, a really intelligent, good child, even though a bit spoiled. They live in a beautiful house in G[enoa] -- in any case, all bodily needs are taken care of. You will also get to know the city and the country. Really, nothing is more important to my aunt than the education of her children. I cannot tell you anything about the salary: you should write me about that since I am ignorant in those matters. You only have to come to terms with my aunt -- the others have no say.

Please write me pretty soon about the particulars, but also with a personal little appendix for me, please.

Kind regards from your

Albert Einstein

52. TO MILEVA MARIĆ

Paradies [Mettmenstetten] Thursday [10? August 1899]

D[ear] D[oxerl]!

Many thanks for your little letter, to which I would have replied by now had I not gone with our landlord on a walking tour in the mountains, which was, by the way, quite delightful (Zug - Einsiedeln - Züricher Obersee). I hope that you received my first little letter, even though it did not contain much of importance, because otherwise you would be certain to bear a bitter grudge against me & think that I am a faithless idler. The vacation is passing in blissful peace and quiet, so that it's studying that represents a change to me, and not loafing, as we are used to from our household. And you, good soul, are writing me that the cramming diet for the examination does you some good, I like that. You are really great & have much vitality & health in your small little body. I returned the Helmholtz volume and am at present studying again in depth Hertz's propagation of electric force. The reason for it was that [I] didn't understand Helmholtz's treatise on the principle of least action in electrodynamics. I am more and more convinced that the electrodynamics of moving bodies, as presented today, is not correct, and that it should be possible to present it in a simpler way. The introduction of the term "ether" into the theories of electricity led to the notion of a medium of whose motion one can speak without being able, I believe, to associate a physical meaning with this statement. I think that the electric forces can be directly defined only for empty space, [which is] also emphasized by Hertz. Further, electric currents will have to be conceived of not as "the vanishing of electric polarization in time" but as motion of true electric masses whose physical reality seems to be confirmed by the electrochemical equivalents. Mathematically, they are then always to be conceived of in the form

$$\frac{\partial X}{\partial x} + \left[\frac{\partial Y}{\partial y} + \frac{\partial Z}{\partial z} \right].$$

Electrodynamics would then be the theory of the motion of moving electricities and magnetisms in empty space: which of the two conceptions must be chosen will have to be revealed by radiation experiments. -- By the way, I had no news so far from Rector Wüst. I will write him shortly.

Here in Paradies it is continuously very lovely, particularly because we have wonderful weather. However, we always have disagreable visits from Mama's acquaintances, whose dull-witted chatter I usually escape by slipping away if we are not just having a meal. Finally, there is also my aunt from Genoa, a veritable monster of arrogance & dull-witted formalism. All the same, I enjoy each and every day of my vacation at this charming and quiet little place. If only you would be again with me a little! We both understand each other's black souls so well & also drinking coffeee & eating sausages etc....

I also almost believe that the whole story regarding you & Mama existed only in my imagination. It probably wouldn't even have occurred to you. Anyhow, write me here the next time. You must have misunderstood me then. I am always delighted with your little notes; send me some again soon.

Affectionate greetings from your

Albert

Don't work too hard!
Regards from Mama & Maja!

My kindest regards to your loved ones.

53. FROM MILEVA MARIĆ

[Kać, after 10 August - before 10 September 1899]

D[ear] Mr. E[instein]

Both letters of yours have found me in our hermitage; thanks a lot for it and please another one very soon again. Surely, you, who have now so much time on your hands, are not going to follow my example just now when I write a little less to you. Every time, your letters remind me so nicely of home. From the succession of our joint experiences a peculiar feeling has formed surreptitiously, which is awakened at the slightest touch, even without the recollection of any particular detail becoming quite conscious, and which makes it seem each time as if I were again in my room.

You are right in not studying hard, if it's only true, it's difficult to totally believe that, better take nice walks now that you have such a good opportunity. All this time I have not gotten further than our garden, we now don't go to town at all, there have been many cases of scarlet fever and diphtheria, so that we prefer staying in our fresh and healthy air. N[eusatz] [Novi Sad] is a rather unhealthy hole, and in addition it is now terribly hot here. The sour-cherry trees are blooming here for the 2d time.

Give my kindest regards to Mrs. Einstein and Miss Maja; I shall be pleased if she writes to me (of course, if you permit it, I don't see why not, except if you really have a good reason). The "cramming" proceeds slowly. Fiedler worries me a lot, this seems to me the hardest nut to crack. You could write me a little on how it goes at the examination, but you must not think that the letter should be filled with the description only. (as a precautionary measure) Do Fiedler and Herzog ask also specific things, examples, or only general questions? And I would like to ask you for one more thing, when you come to Zurich, please leave your notebook on the theory of heat with Mrs. Markwalder, I would like to look up a few things.

You do not mention the date of your departure from Paradies. I'll probably be in Zurich on the 25th, and instead of looking forward to it, I am going there with such mixed feelings, don't you feel sorry for me.

I hope you aren't letting anyone read my letters, you must promise me that: you said you don't like the profane, and if to me this seems profane, you could do that for me! What do you think. The next time I'll use a different mode of address on the letter, I know of a nicer one; it is very pious, but it occurred to me too late. - You must excuse me if my scribbling may betray some absent-mindedness, it seems to me that I contracted a little of this lovely quality, but not for good, I hope.

Accept my kind regards and write very soon to your

D[oxerl]

54. TO MILEVA MARIĆ

Sunday. Paradies [Mettmenstetten, 10 September 1899]

D[ear] D[oxerl]!

I am finally able to write to you after receiving your dear little letter, which I would have liked best of all to answer the very first day. But it had to be forwarded to Aarau, where it is too charming & one always gets pestered in such a lovely way that it is impossible to write.

You poor thing must now be cramming really awfully! If only I could help you a bit, be it merely to bring you some variety, or be it in the studies, or be it as Johann with all the lovely trifles that go with that.

But now this will soon be behind you & then you can again count one more victory in life & you'll get a significant little piece of white paper. I couldn't take the physics notebook to your pad yesterday, when I was in Zurich, because there wasn't enough time for that. But don't sulk because of this, little witch! When you come to Zurich, just come to my pad & take whatever you like & if you cannot find something that should be there, just ask Mrs. Markwalder. (She knows everything?). -- The day after tomorrow I am going with my mother to Milan, and I won't be back in "our" place in Zurich before the start of the semester. I would like so much to try to make the examination time more pleasant for you in Zurich if this wouldn't cause very understandable pain to my parents.

At present I'll be completely bookless for a week, since all libraries are now being inventoried, but within a week I can have the municipal library send me books by Helmholtz Boltzmann & Mach to Milan. But so that you wouldn't frown and worry, I give you my solemn promise that I'll go over everything with you. I think we should once stay in Zurich during vacation time, so as to lead our semester life without lectures in ease and comfort for once, this must really be quite nice.

A good idea occurred to me in Aarau about a way of investigating how the bodies' relative motion with respect to the luminiferous ether affects the velocity of propagation of light in transparent bodies. Also a theory on this matter occurred to me, which seems to me to be highly probable. But enough of that! Your poor little head is full enough of different people's hobby-horses that you had to ride. This being the case, I shouldn't march in mine as well. I don't know what else to say, except that you needn't let this little bit of examination worry you. This is a trifle for you - especially with such harmless competitors.

And now goodbye, don't toil so hard, accept a thousand affectionate regards & use a nicer mode of address the next time you write to your

Albert
Via Bigli 21
Milan

Kind regards to your loved ones
" " from Mama & Maja.

55. TO JULIA NIGGLI

[Mettmenstetten] Monday [11 September 1899]

Dear Miss!

Since everything is already packed, I am using this little piece of paper to enclose a little greeting. I have already given my aunt a plausible explanation about your having accepted another position, and she now has the choice between the governess from Zofingen you recommended and a young woman from Zurich whose name by chance is also Martha Müller. The latter will come to Aunt this afternoon for an interview. At any event, I'll be able to send an answer to your friend, though not a definitive one. In the meantime I had a few verbal battles with the ladies. Tomorrow, then, follows the separation when we return home.

I wish you all the luck in your future position, & most of all, I wish you luck in love, which for you girls is always the most important thing (but for us boys too). Now I'll have to apply myself once again to my domestic duties as the pillar of the housewives & society ladies.

Accept, then, for today my kindest regards and write once more to your

Albert Einstein
Via Bigli 21
Milan

Remember me to your Fritz and your father.

56. TO PAULINE WINTELER

[Mettmenstetten] Monday [11 September 1899]

Dear Mommy!

Since everybody except myself is now busy packing for our departure tomorrow, I happen to have a free hour or so, which seems just made for my writing you another little letter. My heartfelt thanks for all the love you have given me as usual, in spite of the great distress that befell you through my fault. If only the letter in question did some good!

Here I found everything in best order. Instead of a severe lecture, I got only a few maternal looks that could have passed for it. But now everything is forgotten again, as I got used to Aunt's whims, which, after all, are childish, and are in fact only the result of great pampering. One only has to treat such folks properly, and then one can do as one likes & does not give them a headache. I had a very lively conversation with Professor Haab and very animatedly discussed with him the topic suggested by you. In most praiseworthy modesty he took himself as an example & then talked much about pearls that had been cast before swine & thus had found their way to oblivion. He spoke about a hoped-for change, but also about its difficulties. But don't tell *him* anything about that, he would find it very painful that such a young and inexperienced man interferes in his internal affairs. Actually, I would have remained completely passive during the conversation if his amiable nature had not made me open my mouth. Please ask the Professor to give my kindest regards to Mr. Wüst, I think I forgot to do it. In addition to the governess recommended by Miss Julia, Aunt has also in prospect a woman of the same name from Zurich. Both are called Martha Müller. The woman from Zurich is an acquaintance of Mrs. Markwalder & will arrive here this afternoon. In any case, I will see to it that the governess from Zofingen has an answer written to her already today, so that her hands would not be tied.

The reporting of sheer facts has made my letter quite dry, I hope that from Milan I'll be able to have this one followed by a better one, which will then be destined for *you*.

Pauli should not be angry about the postcard from my sister. She wouldn't have allowed herself this teasing had I not been behind it.
Cordially, your

Albert.

Kindest regards to your loved ones & also Miss Julia!

57. TO MILEVA MARIĆ

Milan, Thursday. [28? September 1899]

D[ear] D[oxerl]!

You are such a splendid girl for writing me so nicely despite all your hard and taxing work. But you should also know that your letters always make me so happy that everybody teases me for it. You must have swallowed a lot of book dust, poor thing, but soon it will be behind you - I really feel with you. I too have done much bookworming & puzzling out, which was in part very interesting. I also wrote to Professor Wien in Aachen about the paper on the relative motion of the luminiferous ether against ponderable matter, which the "Principal" treated in such a stepmotherly fashion. I read a very interesting paper published by this man on the same topic in 1898. He will write me via Polytechnikum (when it's for certain!). If you see there a letter addressed to me, you may take it and open it.

I'll be back "home" about the 15th. I look forward to it with great joy, because our place is still the nicest and coziest after all. Maya is going to Aarau chiefly because we know a family there very well and because life is less expensive there; we must keep that very much in mind.

Mrs. Markwalder feels herself bound to.... She is an angel of presentiment. I have already written to her that I agree & have given her permission to assign the room in whichever way. I would move to somewhere in the Plattenstrasse, but not into your house - for the sake of people's tongues. I would move to the Zürichberg if it would not be so far away from "us."

By the time you receive my letter, the Fiedlering will probably be already behind you - I think so much about it. Everything will go all right - your hard little head guarantees me that. If only I could peep through the keyhole! When one is taking such an examination, one feels so responsible for everything one thinks and does as if one were in a penal institution. Isn't it so? At this time I was laughing so much with Grossmann about these things - but an uninvolved party could have commented, "laughing on the outside, cracking in the inside."

My sister will probably not stop in Zurich at all; rather, I shall accompany her to Aarau, but I shall not stay there beyond that. Here I again feel rather uncomfortable, because the climate does not agree with me at all & because of the lack of some definite work I brood too much - in brief, I see & feel that I am not under your beneficient thumb, which otherwise keeps me within bounds.

Neuweiler is the black one that always drinks milk. It also has eyeglasses and very little fat. You don't have to be afraid at all that I'll be going to Aarau very often now. Because the critical little daughter is coming home, with whom I fell so terribly in love 4 years ago. To be sure, I feel otherwise very safe in my high castle

Peace of Mind. But if I saw the girl again a few times, I would surely get crazy again, I am aware of that & fear it like fire.
Once I am back in Zurich, we shall immediately climb the Ütliberg. There we can have fun unpacking our memories of Säntis; I picture it again to myself in such gay colors. And then we will start immediately with Helmholtz's electromagnetic theory of light, which 1) out of fear 2) because I did not have it, I still have not read.
Thousand affectionate greetings from your

Albert

58. TO MILEVA MARIĆ

Milan, Tuesday [10 October 1899]

D[ear] S[weet?] D[oxerl]!
Now, you are a fine one! It's already the 4th day that she has been sitting very cozily at the examination & has not yet uttered a single word to her good colleague and coffee-guzzling pal. Isn't that horrible? I shall compose a fire-and-brimstone sermon and give it to you in person next Monday, and early in the morning at that. And if the maid says that you have left & I see your polished little boots in front of the door, which seems to be happening from time to time - then I'll simply wait a little or I'll get a shave.
Sunday I am taking my sister to Aarau & the very same day I'll show up in Zurich at the place of my beloved ex-landlady, who simply has not responded to a postcard in which I dared to ask her whether it is within her "inscrutable providence" to billet me somewhere else. Thus I, poor postal parcel, must wait until I get enlightened about the place of my destination. - If I think how you are now buried in work, my anger caused by your non-writing melts away like wax. You poor creature, you in fact have it much harder than I had it the last year because you are so alone! But stop - I already see you smile over my consolation job & think: Such a matter is of little concern to a Dockerl; it knows itself what it wants and can & has it already demonstrated several times.
But now something nicer - I have in mind our household, of course. It will again become lovely there. I am bringing a few marvelous delicacies from Mama, who has also promised to send us frequently something to the household: directly to Plattenstr. 50. In the meantime, do get Helmholtz's electromagnetic theory of light! I have already developed quite a hunger for it.
I have been studying here a lot & have completed my considerations about the study of the basic laws of thermoelectricity. I have also devised a method of great simplicity, which permits one to decide whether the latent heat in metals is to be reduced to the motion of ponderable matter or of electricity, i.e., whether an electrically charged body has a different specific heat than an uncharged one. All these questions are connected with the analysis of the thermoelement. The methods are very simple to carry out & do not require any equipment that is not readily available to us.
But that's enough for today, otherwise my old folks are going to tease me for my writing so much without a reply. Cordial greetings & a happy reunion! Your

Albert

59. MUNICIPAL CERTIFICATE OF RESIDENCE AND GOOD CONDUCT

Police Department of the city of Zurich

Zurich, 18 October 1899

CERTIFICATE OF RESIDENCE AND GOOD CONDUCT

This is to attest that Mr. Albert Einstein, born on 14 March 1879, from Ulm, Württemberg, stud. math., residing at 4 Unionstr. in Zurich V, maintains his actual and uninterrupted residence in this city since 29 October 1896 and, as far as it is known, has an umblemished reputation.

The Police Chief
E. H. Müller
The Chief of the Central Control Bureau:
Bühler

Contr. No. 3599
Fee 60 Rpn.

60. TO THE SWISS FEDERAL COUNCIL

Zurich, 19 October 1899

To the High Federal Council of the Swiss Confederation in Bern

Highly honored Mr. Federal President!
Highly honored Messrs Federal Councillors!

The undersigned, Albert EINSTEIN, born on 14 March 1879, originally from Ulm, Württemberg, student of mathematics at the Polytechnicum in Zurich, submits hereby to you his humble request that you grant him the approval for the acquisition of the Swiss cantonal & municipal citizenship.

In support of my request I am taking the liberty of enclosing the following documents:

1.) A certificate of residence and good conduct by the City of Zurich regarding my actual and uninterrupted residence in this city since 29 October 1896.

2.) The certificate of release from Württemberg citizenship.

Recommending my request to your most benevolent consideration, [I] remain with the most profound respect!

Albert Einstein
4 Unionsstrasse Zurich-Hottingen

2 enclosures!

61. FROM MILEVA MARIĆ

[1900?]

My dear Johonesl!

Since I like you so much and you are so far away that I cannot give you a kiss, I am now writing this little letter and am asking you whether you like me as much as I do you? Answer me *immediately*. Thousand kisses from your

D[oxerl]

[This letter is written humorously, in pseudo-dialect.]

62. TO THE SWISS DEPARTMENT OF FOREIGN AFFAIRS

Zurich, 28 February 1900

To the High Political Department of the Swiss Confederation, Bern

Highly honored Mr. Federal Councilor!

I am taking the liberty to transmit to you, enclosed, my father's written authorization regarding my acquisition of Swiss cantonal and municipal citizenship, and am using this opportunity to respectfully request an early and favorable decision on my application.
With the deepest respect!

Albert Einstein, stud. jur.

1 enclosure.

63. MILEVA MARIĆ TO HELENE KAUFLER

Zurich, 9/III 1900

My dearest Miss Kaufler!

[...]

Yesterday E. celebrated his birthday and today were to come the afterpleasures. Mrs. E. has sent a lot of the very sweetest delicacies; and you should have seen what a magnificent effect this had on her son: he passed beaming through Plattenstr. with the box in both hands and was so pleased that he did not look at anybody. [...]

[...]

Prof. Weber has accepted my proposal for the diploma thesis, and was even very satisfied with it. I am very happy about the investigations I'll have to do for it. E. has also chosen a very interesting topic. - He sends his best regards to you and Miss Ida.

[...]

Mileva Marić

[...]

64. MILEVA MARIĆ TO HELENE KAUFLER

Zurich, Monday [4 June - 23 July 1900]

My dear, dear Miss Kaufler!

[...]

Also, I was rather anxiously waiting for certain reports, but alas! these do not seem to be very favorable for me, because if even you, you dear, kind-hearted person, stoop to mockery, then the situation must be really bad. Do you think that she does not like me at all? Did she make fun of me really badly? You know, I seemed to myself so wretched at the moment, so thoroughly wretched, but then I comforted myself all the same, because after all the most important person is of a different opinion, and when he paints beautiful pictures of the future, then I forget all my wretchedness, or do you think that I shouldn't?

[...]

For the time being, everything goes on as usual in our Zurich, but the Pension Engelbrecht seems to stand before great changes: among others, Miss Drazic and Miss Botta also want to bid farewell to it; heavy blows of fate for Miss Engelbrecht! The girls seem to be somewhat cross with me too, though I have not the slightest idea why; maybe I must even atone for other people's sins, or God knows what. [...] Today Mr. Einstein also made up a little satirical poem about them, very good but very wicked, and he will give it to them. This will really be something. --

Mr. Einstein sends you his best regards, he is extraordinarily pleased that you liked his "old lady" so much, and almost seems a little envious because his father is said to be so handsome.

[...]

Mileva Marić

65. TO ZURICH CITY COUNCIL

Zurich, 26 June 1900

City Council Zurich

Highly esteemed Mr. City President!
Highly esteemed Messrs. City Councilors!

The undersigned, ALBERT EINSTEIN, stud. math. at the Federal Polytechnikum, born on 14 March 1879, from Ulm, Württemberg, herewith respectfully requests that on payment of the legal fee you admit him to the Citizens' Federation of the City of Zurich.

Domiciled in Zurich since 29 October 1896, I have always devoted myself to studies and believe that in no manner have I given grounds for any conclusions that could be an obstacle to the granting of the citizenship.

I take the liberty of attaching to my request the approval of the Federal Council, required by municipal law for the acquisition of

Swiss cantonal and municipal citizenship, and will gladly provide any further information.

With profound respect!

Albert Einstein
Unionstrasse 4, Zurich V.

1 enclosure

hopes to become assist. physics, decision w[ill be brought] by end of July, but intends to stay in Zurich in any case, if possible. The aforementioned position cancelled, will stay nevertheless. Teacher, later docent. Passed the diploma examination. ***to wait until mid-September.***

66. MUNICIPAL POLICE DETECTIVE'S REPORT

Municipal Police Zurich District V

Zurich V, 4 July 1900

To the Police Inspectorate Zurich

Inquiries made about EINSTEIN Albert, stud. math., born 1879, from Ulm, now residing at Hägi's, Unionstrassse No 4, showed that he is a very eager, industrious and extremely solid man. (Teetotaler) He is said to visit every week with Professor Stern, Engl. Viertelstrasse No. 58, and with Fleischmann, of the firm Fleischmann & Comp. Grain Trade, Bahnhofstrasse 65, to both of which places he is from time to time invited.

Einstein is single and his parents are said to live in Milan. In Klosbachstrasse, where Einstein also lived for several weeks, they have given a [character report] on him which is excellent in every respect.

Hedinger Det[ective]

67. ADOLF HURWITZ TO HERMANN BLEULER

Zurich, 27 July 1900
Pension Sonnenberg

Mr. H. Bleuler
President of the School Council
Here

Highly honored Mr. President!

This year's final diploma examination in Department VI A produced the results summarized in the table below:

	Theory of Functions	Geometry	Arithm.& Algebra	Theor. Physics
Ehrat	*11*	*11*	*4½*	*5*
Grossmann	*11*	*12*	*4*	*4½*
Kollros	*12*	*11*	*4½*	*4½*

Astronomy	Diploma Thesis	Grade Total	Average Grade
5	*20*	*56.5*	*5.14*
4	*22*	*57.5*	*5.23*
6	*22*	*60*	*5.45*

	Theor. Physics	Exper. Physics	Theory of Functions	Astronomy
Einstein	*10*	*10*	*11*	*5*
Marić	*9*	*10*	*5*	*4*

Diploma Thesis	Grade Total	Average Grade
18	*54*	*4.91*
16	*44*	*4.00*

Based on these results, the Conference of Examiners moves that diplomas be granted to candidates Ehrat, Grossmann, Kollros, and Einstein, but not to Miss Marić. Respectfully

Prof. Dr. A. Hurwitz

Motion approved:
28 July 1900 H. Bleuler

68. TO MILEVA MARIĆ

[Melchtal] Sunday morning [29? July 1900]

My dearest Doxerl!

Because I am writing in my bed
This just won't be so terribly neat!
But never mind, just go on scrawling,
For Doxerl is interested in it all the same!...

So the day before yesterday I arrived, as planned, with the Terrible Aunt at Sarnen, where Mama, Maya, and a carriage were waiting for us. Thereupon I got smothered with kisses. Then we drove off; but soon Maya and I got off to stroll a little bit. On that occasion Maya told me that she did not dare to report anything about the "Dockerl affair," and she also asked me to "spare" mama - which means - not to blurt out everything.

We come home, I into Mama's room (just the two of us). First I have to tell her about the examination, then she asks me quite innocently: "So, what will become of Dockerl?" "My wife," say I, equally innocently, but prepared for a real "scene." This then ensued immediately. Mama threw herself on the bed, buried her head in the pillow, and cried like a child. After she had recovered from the initial shock, she immediately switched to a desperate offensive, "You are ruining your future and blocking your path through life". "That woman cannot gain entrance to a decent family." "If she gets a child, you'll be in a pretty mess." At this last outburst, which had been preceded by several others, my patience finally gave out. I rejected the suspicion that we had been living in sin with all my might, scolded properly & was just ready to leave the room, when Mama's friend Mrs. Bär entered the room, a small, lively little woman full of life, such a sort of hen of the nicest kind. Thereupon we immediately started to talk with the greatest eagerness about the weather, new spa guests, ill-behaved children, etc. Then we went to eat, after that we played some music. When we said "good night" to each other in private, the same story started again, but "piu piano." Next day things were already better, and this, as she herself said, for the following reason: "If they have not yet had intimate relations (so much dreaded by her) and will wait so long, then ways and means will surely be found." Only what is most terrible for her is that we want to stay together forever. Her attempts at converting me were based on speeches like: "She is a book like you - but you ought to have a wife." "When you'll be 30, she'll be an old hag," etc. But as she sees that in the meanwhile she is accomplishing nothing except to make me angry, she has given up the "treatment" for the time being.

Life and people here are hopelessly dull & I perfectly understand the dissatisfaction of Maya, who is again mad with joy as she looks forward to Aarau. Each meal lasts 1 hour and more; you can imagine what a hellish torture [that is] to me. Since, furthermore, the weather is bad, in my desperation I fled to Kirchhoff. Besides those already mentioned, our constant hangers-on include my aunt, the Englishwoman, the contessa with her daughter, who is as beautiful as she is stupid and cold. For Mama's sake I must flatter all of them & play music - otherwise she is offended, all the more so as she is doubly sensitive because of the affair.

If only I could be soon again with you in Zurich, my little sweetheart! Thousand greetings and colossal kisses from your

Johannesl

"Kiss Maya"
Don't write anymore to our d[ear] Weber, he is going to be in the country.

69. TO MILEVA MARIĆ

[Melchtal] Wednesday evening [1 August 1900]

My sweet little one!

How glad I am to know that by now you are home with your good old lady, who is again marvelously fattening up my little Doxerl, so that she will rest again in my arms plump as a dumpling and healthy and cheerful, one fancies - in the beloved oxistant's arms; even though, in fact, I have no news yet from Zurich, the high spirits produced by the carefree life and good food are giving me confidence. I think I have not been able to kiss you for a whole month and I thoroughly long after you. Such a nimble, active thing like my Doxerl with her dexterous hands cannot be found in the entire ant hill of the hotel. The mama-in-law has already made peace with me, more or less & is gradually getting reconciled to the unavoidable. She has already regained her good cheer. I have also written to Papa; he announced a separate letter to me; he too will certainly resist at first, but all this is of no importance.

I long terribly for a letter from my beloved witch. I can hardly grasp that we will be separated for so much longer - only now do I see how frightfully much I love you! Indulge yourself thoroughly, so that you'll become a blossoming little sweetheart and wild like a street urchin.

Melchthal is a wonderful little stream valley formed by mountains that are tall but not covered by glaciers. Our hotel in particular is an outstanding feeding establishment; but I do not feel comfortable in this loafing among these people gone soft. Especially when I see the decked-out lazy women who are always disgruntled about something, I think proudly : Johonnesl, your Doxerl is a maiden of a totally different kind. Brandenberger is also here together with *bride*, a young woman from Zurich whom I like very much. The two are visibly in a state of bliss - a very lovely little couple.

I was yesterday with Maya on a rather high mountain, where we found many edelweiss. We had a marvelous view, especially of the gigantic firn fields of the Titlis.

Toward mid-August we shall go to Papa in Italy, to spend also some time in a more southerly location. Before that, I shall also go to Zurich to see about my job. I have still not heard from Ehrat either. As it rains much, I have already studied a lot, mainly the notorious investigations on the motion of a rigid body by Kirchhoff. I marvel at this great piece of work again and again. My nerves have calmed down so much that I again study with bliss. How are yours doing?

Greetings to your family! Most heartfelt kisses from your

Albert

70. TO MILEVA MARIĆ

[Melchtal] Monday [6 August 1900]

My dear little one!

Your first dear little letter from the homeland arrived yesterday First I read your letter in the quiet little chamber, then twice more & then for a long time I kept reading between the lines with great delight & then I slid it grinning in the pocket. The "Mama-in-Law["] is very congenial & and does not mention the "delicate subject," all the more so because my happy cheerful mood, my popularity among the guests & my "musical successes" are balm on her wounded mother-in-law's heart, so that the atmosphere is by now outright cozy.

But our correspondence, dear sweetheart, seems to be under an evil spell, seeing that you had not received my letter at the time you mailed yours. This is the 3d one I am sending you.

I have not yet received a report from Zurich. I assume I will have to look out for my matter myself. In view of Ehrat's conscientiousness, the only thing I can think of is that his matter is still up in the air.

Take a thorough rest, sweetheart, you can do enough wonderful studying with your Johonnesl later on. But now rest and enjoy your carefree life.

Papa has now also written me a sermonizing letter for the time being, but he promised me that the main thing will follow orally, to which I am most dutifully looking forward to. I can understand my old folks quite well. They consider a wife as the man's luxury, which he can only allow himself once he makes a comfortable living. However, I have a very low opinion of such an attitude toward the relationship between man and woman because according to it, a wife and a whore differ only insofar as the former is able to extort a contract for life from the husband, due to her more favorable life circumstances. Such a view is the natural consequence of the fact that with my parents, as with most people, the senses exert a direct mastery over the feelings, while with us, by virtue of the happy circumstances under which we are living, the enjoyment of life is infinitely broadened. But we must not forget how many existences of the first kind it takes to make this possible for us; because in the social evolution of humankind the former are a far more important constituent. Hunger and love are and remain such important mainsprings of life that almost everything can be explained by them, neglecting the other leitmotifs. I am therefore trying to spare my parents, without giving up anything I consider to be good - and that is you, my dear sweetheart!

If you haven't yet told anything to your family, don't do it! I think that this is better for all parties concerned. Otherwise they might start having the same unnecessary worries and qualms as my family. But after all, you are wise and know them, and know better what you have to do.

When I do not have you, I feel as if I were not whole. When I am sitting, I would like to walk; when I am walking, I look forward to being at home, when I amuse myself, I would like to study, and when I am studying, I feel a lack of contemplativeness and repose & when I go to bed, I am not satisfied with how I passed the day.

Be cheerful, dear sweetheart. Kissing you tenderly, your

Albert

71. TO MILEVA MARIĆ

Zurich Thursday [9? August 1900]

My dear sweetheart!

You are surprised, aren't you, that I have popped up here again! But I am using the first possible excuse to get out from the boring environment, even though my mother made it her duty to observe the deepest silence about the "affair." She behaved as if nothing had happened, handed me your letters in person, did not notice when I was writing to you - to put it briefly, she gave up the open battle & probably will only fire off the philistine cannons jointly with Papa. The latter promised me in his last letter that he will visit Venice with me, because our power plants are not far from there. I would also like to get initiated a little into administration so that I could take Papa's place in case of emergency. He too does not mention you any longer. I would have done better, sweetheart, had we kept everything to ourselves and had I put off the old folks. But no harm has been done, my dear sweetheart, Papa and Mama are very phlegmatic people and have less stubbornness in their whole body than I have in my little finger.

Just as my old Zurich makes me feel at home, so do I miss you, my dear little "right hand." I can go wherever I want - but I do not belong anywhere an[d] I miss the two little arms and the glowing little girl full of tenderness and kisses. How sorry I felt for the Catholic clergymen who were in Melchthal! The measurements of my tender little feet I'll send you another time, now don't you start wriggling again. In return, however, you'll get a big kiss for your hen-like eagerness! - But now, as to the *excuse*. On the evening of the day before yesterday I received a postcard from Ehrat, in which he writes that he proposed me for a temporary position with the insurance office in which he is working at present. One gets 8 fr per day & has 8 hours of idiotic drudgery. But I declined, believing that I could use the vacation better. One must shun such stultifying affairs. I am remaining here to see how the matter will develop and finally to straighten out my "business and political" affairs. It seems that Ehrat's competition also includes Matter, whose election would yield me the position with Hurwitz. So, courage, little witch! I can't wait for the moment when I'll be able to hug you and press you and live with you again. And cheerfully we will go at it and work and have heaps of money. And if the next spring is beautiful, we will get flowers in Melchthal.

Tender kisses from your

Albert

72. TO MILEVA MARIĆ

Zurich Tuesday [14? August 1900]

Dearest little sweetheart!

Once again a few lazy and dull days flitted past my sleepy eyes, you know, such days on which one gets up late because one cannot think of anything proper to do, then goes out until the room has been made up, then studies for a few hours until one gets too tired. Then one hangs around and looks halfheartedly forward to the meal, languidly meditating about highly important philosophical questions & whistling a little while doing so...... How was I able to live alone before, my little everything. Without you I lack self-confidence, pleasure in work, pleasure in living - in short, without you my life is no life.

I was even making visits in order to distract myself. For example, I called on Mrs. Markwalder, who still shows the same apathetic-languid amiability & sees everything in an undefinable daze; how lucky that I do not lodge any longer with her. I have also visited Jungferli [the little virgin], who is still one of the nicest and liveliest persons we know here. She is now leaving for good; she is moving to a small town in the Thurgau. I also visited your landlady. She said that your suitcase had gone a long time ago. She asked whether you perhaps want to keep the room, she would then make a special arrangement. But I turned this down (the tyrant! you'll probably think).

But I will not grant you leave beyond the first days of October;

that is just long enough. I am leaving Saturday for Italy to partake at my father's in the pleasures of the "holy sacrament"; but the brave Swabian is not afraid. I hope I will not become so moldy when I grow old, then it'll be alright. Your old folks are proof that one can also be different -- they must be splendid people. Just don't tell them too much about me, otherwise they too might get scared. Had I been smarter, I would have kept my mouth shut. Why did I not take better to heart my motto omnes tractandi sunt? But never mind, this will make it all the more pleasant when we will again have each other in Zurich and will enlarge our wisdom while having some fragrant coffee! That your mommy feeds you well and your sister teases you mightily is alright & that you are longing for me makes me proud!

Don't study hard when your books have arrived, get rested instead, so you'll become again the old street urchin. There is only a single thing that I do wish & request from you, and that is that you feel well. But if this is not the case, I'll spank you.

Ehrat still does not have the position in Frauenfeld. He is in competition with Matter. But in any case, one of the two will get it. Thus, I am taken care of under all circumstances. I could have had a position in life insurance for 3 weeks for 8 fr. per day, but I declined because I thought that I could use my vacations better by studying something proper & then in Italy learning the trade of my father. After all, it could happen that he suddenly fell ill or would be otherwise engaged & he has nobody at his disposal. How lovely it will be next year!

Affectionate greetings and kisses, especially the latter, from your

Albert

73. TO MILEVA MARIĆ

Milan Monday [20 August 1900]

My dear little one!

Schnadahüpfl [a humorous improvised four-line dialect verse, literally tailor-hop]:

O my! the Johonzel,
He is totally crazy.
He was thinking about his Doxerl
And he squeezed his pillow.

When my sweetheart is sulking,
I become soft as a rag,
But she only shrugs her shoulders
And says: I don't care.

My old folks they think,
This is a stupid thing...
But they don't say anything,
Lest they'd get whacked over their head.

The little beak of my Doxerl
I would like to hear it
And after that cheerfully
Shut it off with mine...

So, sweetheart, it's already two days that I am here with my old folks and quite pleased with them. There is no "treatment" in sight. I have never talked about you in particular, but I have often dropped your name here and there. As I see it, they have nothing against our relationship, evidently because they do not think any longer that we are ruining our future. Besides, they also know that I won't let myself be influenced. If I don't provoke them, everything will take its jolly course - we a cheerful lively couple and they satisfied and pleased with that.

Oh, how happy I will be when I can clasp you to my heart again! This will be in the first days of October! But now you should have it nice, my only sweet little woman. I haven't yet heard anything about the "position." But I am taking it easy. If I don't get any, the "whole family" will simply have to give private lessons. Thanks to the good domestic feed and the cheerful disposition of my parents, my optimism has again grown immensely. My father has become a completely different man now that he no longer has to worry about money. That all the dark clouds have vanished you can also see from the fact that he is taking a trip to Venice with me after we have visited his power plants together. I would like to kiss you out of pleasure and delight, my little dear angel.

But you have not written to me for a long time, you wild witch! Are you afraid it will "miss its target" or are you furious, you little rascal? Or do you just want to make me curious and hungry? Or can it be that you are afraid of sisterly jokes?

The child makes herself hard to get
What does he think about that?
It is devoted, after all, with all its fibers
To its lad.

But because you are such a wild little rascal, I am stopping now, furious like the devil!

Greetings and kisses from your

Albert

74. TO MILEVA MARIĆ

Milan via Bigli 21. Thursday night in bed
[30 August or 6 Sept 1900]

My dear Miez!

Today I received a registered letter from you, by which I can tell that you fear it may fall into somebody else's hands. No, sweetheart, I got all your letters on time, and also the money long ago in Melchthal. You can always write to me exactly the way you feel, because for my parents to take away a letter would be as unwise as it is useless. By the way, you don't have to be afraid of something like that, if for no other reason than because I am sure that my parents are incapable of acting in such a way. I have put Mama to the test. My parents are very distressed about my love for you, Mama often cries bitterly & I am not given a single undisturbed moment here. My parents mourn for me almost as if I had died. Again and again they wail to me that I brought disaster upon myself by my promise to you, that they believe that you are not healthy........ oh, Doxerl, this is enough to drive one crazy! You wouldn't believe.how I suffer when I see how both of them love me and are so disconsolate as if I had committed the greatest crime & not done what my heart and my conscience irresistibly prompted me to do. If only they would know you! But they are as if bewitched & think that this is what I am. On Saturday I am going on the trip with Papa, to Venice too. I was so sad, that I didn't want to go with him; but this alarmed them so terribly that I got quite scared.

I'll only be able to recover from this vacation gradually in your arms - there are worse things in life than an exam. Now I know it. This is worse than the difficulties in the world.

My only distraction is studying, which I am now doing with twice as much love & and my only hope is you, my dear faithful soul. Without the thought of you I would not want to live any longer in this sorry human crowd. But possessing you makes me proud & your love makes me happy. I will be doubly happy when I can press you to my heart again and see your loving eyes which shine for me alone, and kiss your dear mouth, which trembled in bliss for me alone.

Thank God that August crept past. Only 4 more weeks, and then we are united again and can live to please each other. But then I'll not let you go so soon again!

I am spending many evenings here at Michele's. I like him very much because of his sharp mind and his simplicity, & also Anna and, especially, the little brat. His house is simple and cozy, even though the details show some lack of taste.

Kissing you from the bottom of his heart, your

Sweetheart

Friday. So, tomorrow we are leaving for the trip, but we will be back in a week, so just keep sending me your little letters. Luigi A[nsbacher] might come to visit us.

For the investigation of the Thomson effect I have again resorted to another method, which has some similarities with yours for the determination of the dep[endence] of κ on T & which indeed presupposes such an investigation. If only we could already start tomorrow! With Weber we must try to get on good terms at all costs, because his laboratory is the best and the best equipped.

Kissing you tenderly, your

Albert

How is your little neck?

I am investigating the following interesting problem for Michele: How does the radiation of electric energy into space take place in the case of a sinuisoidal alternate current? About the amplitude of the waves produced as a function of the frequency of vibration, etc.

75. TO MILEVA MARIĆ

[Milan] Thursday. [13? September 1900]

My dearest Doxerlin!

3/4 of the stupid time is now over, soon I'll be again with my sweetheart and kiss it, hug it, brew coffee, scold work laugh stroll chat...+ ad infinit.! This will again be a cheerful year, will it not? I have already declared that I am staying with you for Christmas. I cannot wait to have you again, my all, my little beast, my street urchin, my little brat. Now that I think of you, I just believe that I do not want to make you angry & tease you ever again, but want to be always like an angel! Oh, lovely illusion! But you love me, don't you, even if I am again the old scoundrel full of whims and devilries, and as moody as always!

I don't know whether I have been writing you these days as regularly as usual. But don't make angry faces because of that - my aunt is visiting here (the praised one from Genoa) with her little daughter, a sadly spoiled brat. So I have no room whatever where I can be alone to write to you. But if I write in front of my parents, they think that I am doing it to spite them. By the way, they are very nice to me, especially Papa; they seem to have reconciled themselves to the inevitable. Both of them will be very fond of you once they get to know you. Now I am very glad that I told everything. They should be glad too, because nowhere in the world would I bee able to find a better one than you, now, when I see other people, I recognize this more than ever. But I also value you & love you the way you deserve. Even my work seems to me pointless and unnecessary if I am not telling myself that you are happy with what I am and what I do. And I am finally sending you the sketch of my gigantic little foot, which I so often forgot to send you.

Johannsel's foot! [accompanied by a sketch]

Since you have such a huge imagination & are used to astronomical distances, I believe that the accompanying work of art will suffice.

I am glad that you stroll around so much & have gotten a good sunburn - how am I going to press my little Negro girl to me! I am also looking forward very much to our new studies. You must now continue with your investigation - how proud I will be when maybe I'll have a little doctor for a sweetheart while I am myself still a totally ordinary man.

But I already mentioned in my last letter, didn't I, that Matter got the position in Frauenfeld & that consequently I shall probably advance to Hurwitz's servant with God's help (of course because he is a man). Doesn't matter, Toxerlin, at least it's yours.

The Boltzmann is magnificent. I have almost finished it. He is a masterly expounder. I am firmly convinced that the principles of the theory are right, which means that I am convinced that in the case of gases we are really dealing with discrete mass points of definite finite size, which are moving according to certain conditions. Boltzmann very correctly emphasizes that the hypothetical forces between the molecules are not an essential component of the theory, as the whole energy is of the kinetic kind. This is a step forward in the dynamic explanation of physical phenomena. Do you also know already that for some time I have been shaving myself, and that with great success? You'll see, Toxerline. I can always do it while you are brewing the coffee after lunch, so as not to continue studying as

I usually do, while the poor Doxerl of course has to cook and the lazy Johanzel never budges once he has quickly carried out the order "grind this."

Heartfelt greetings & kisses, my dear sweetheart & my kindest regards to your loved ones from your

Albert

76. TO MILEVA MARIĆ

Zurich [Milan] Wednesday evening in bed. [19 September 1900]

My dear Doxerl!

Thank you for your kind little letter with the nice dreams of the future, the noodles & the Xantippe-ing, and the plan to bring along your fat little sister into our "European culture." So that she would get a truly high opinion of it & be really impressed by us, I have already bought two little coffee spoons for our household. How lovely it will be again when I'll be able to press you to myself again, my little street urchin, my little verandah, my everything!

Imagine, tomorrow morning I am again running away to the mountains - I'll be climbing a mountain near Lago Maggiore & I shall then visit Isola Bella. How wonderful it would be if Madame Federico Maier could also be there to take a look at all that splendor and then, during the quiet evening, sweetly and nicely chase away *his* peculiar notions, don't you think so, sweet Miez! How I am going to bite you and hug you once I'll have you again - and now it will have to wait more than 3 weeks because of the stupid little goiter. How are things in that respect ? On 1 October I am probably going to Zurich, to talk personally with Hurwitz about the position. This is better than writing. I should look around for sources of income for you? I think I will look around for private lessons, which you possibly could take over. Or do you have also something else in mind? Write me about that!

However things turn out, we are getting the most delightful life from [in?] the world. Beautiful work and together - and what's more, we both are now our own bosses & are standing on our own feet & can enjoy our youth to the hilt. Who could have it better? And when we have saved enough money, we shall buy bikes and go on a biking trip every few weeks. Your dear sister, whom I already know through her lively little letters, will surely like it here with us - I do not have to tell you that she is most welcome, the cheerful, obstinate thing. In a playful moment I wrote yesterday a letter to a former teacher of mine in Munich, of whom I was especially fond, I'll see whether he is going to answer it.

By now I have studied the entire Boltzmann and a part of spherical harmonics, in which I have now even got quite interested. Beggars can't be choosers...

Luigi A[nsbacher], about whom we are always teasing Maja, is going to arrive in a few days - I am looking forward to playing music with him.

I have also thought about the fact that my little sweetheart will be homeless when she comes to Zurich. Unfortunately, Mrs. Hägi no longer has room in her new apartment. But I will look around for something. Perhaps I might just as well hunt for a room for the two of you. If only I were not afraid of the responsibility - because my little brat is very capricious.

Poor little Helen has now fallen for him after all, due to his laudable persistence - now her fine spirit will suffocate in his fat - a sad psychological prophecy. It's really a pity for her. Moreover, I think that in a short time he will be the same scoundrel he used to be. Something like that doesn't disappear easily.

The "groom" has thus become a real "husband"? You see, real miracles do happen in our skeptical age too!

Friendly greetings to your family! Kissing you all over, wherever you allow it, your

Johanzel

77. TO ADOLF HURWITZ

Milan, 30 September [1900]

Esteemed Herr Professor!

My friend Ehrat has written to me that Dr. Matter, who has been your assistant until now, has obtained a position as a Gymnasium teacher in Frauenfeld. I am therefore taking the liberty to inquire respectfully whether I have a chance of becoming your assistant.

I would not have taken the liberty of troubling you with such an inquiry during vacations were it not for the fact that the granting of citizenship in Zurich, for which I have applied, has been made conditional upon my proving that I have a permanent job.

Thanking you in advance for your kind response, I remain respectfully yours

Albert Einstein
Via Bigli 21
Milan

78. TO ADOLF HURWITZ

Milan, 26 September [1900]

Esteemed Herr Professor!

Thank you for your kind letter. It made me very happy to learn that I have a chance of getting the position. Since lack of time prevented my taking part in the mathematics seminar, and no opportunity was offered for seminar exercises in theoretical and experimental physics, there is nothing in my favor except the fact that I attended most of the lectures offered. I think I should therefore also mention that I occupied myself mainly with analytical mechanics and theoretical physics during my university years.

I remain respectfully yours

Albert Einstein

79. TO MILEVA MARIĆ

Milan. Wednesday. [3 October 1900]

Dear Doxerl!

Soon I will really have a bad conscience for not having written to you frequently enough lately, even though I wouldn't be able to say exactly when it was that I wrote to you the last time. In fact, I shouldn't make you angry anymore, now that we are to see each other again so soon, because of the scolding, but he is just dumb, your Johannsel.

I am glad that your sister will be coming after all. We will cure her homesickness & other whims, even though I cannot do it in Serbian. So, the sledging has made the things turn out all right. There is nothing like a female! (Of course, as a little natural scientist, you are always excepted in such considerations.)

I have now prolonged my stay here until Sunday morning, because now I feel quite good here. Even though with hesitation and grudgingly, my old folks withdrew from the fight about Dockerl when they saw that they must lose. Now they rejoice in the good weather & save me from any further debates. Hurwitz has not yet written anything further, but I have hardly any doubts.

Michele has already noticed that I like you, because, even though I didn't tell him almost anything about you, he said, when I told him that I must now go to Zurich again: "He surely wants to go to his [woman] colleague, what else would draw him to Zurich?" I replied, "But unfortunately she is not yet there." I prodded him very much to become a Dozent [university lecturer], but I doubt very much that he'll do it. He simply doesn't want to let himself and his family be supported by his father, this is after all quite natural. What a waste of his truly outstanding intelligence.

In physical chemistry I am now quite well versed. I am delighted by the accomplishments attained in this field in the last 30 years. You will enjoy it when we'll go over it together. The physical methods of investigation employed are also very interesting. The grandest of all is the theory of ions, which has splendidly proved itself in the most diverse areas.

The results on capillarity, which I recently found in Zurich, seem to be totally new despite their simplicity. When we come to Zurich, we shall seek to get empirical material on the subject through Kleiner. If a law of nature emerges from this, we will send it to Wiedemann's Annalen.

At present the ex-groom Fritz Winteler is visiting with Anna; he is a disgusting shop-talker & will again be Assistent in Darmstadt.

You too don't like the philistine life any longer, don't you?! He who tasted freedom cannot stand the chains any longer. How lucky I am to have found in you a creature who is my equal, who is as strong and independent as I am myself! Except with you, I feel alone with everybody.

Many affectionate kisses from your

Albert

Regards to your loved ones!

Dolderstr. 17.

80. MILEVA MARIĆ TO HELENE KAUFLER

Katy near Neusatz [Before 9 October 1900]

My dearest Helenchen!

[...]

I'll come again to Zurich for the winter and so will Albert; we look forward to seeing each other with boundless happiness. He writes very often and is very cheerful as usual. He always sends you his regards and asks how you and your sisters are doing. Do you still remember his prophecy? Now you will soon have the opportunity to refute his blasphemous words through facts. Such a boisterous fellow, and yet I love him so much! But you, dearest, you love both of us because of that, I know it.

[...]

Miza

81. TO HELENE KAUFLER

Zurich Thursday [11 October 1900]

Dear Miss Kaufler!

So, it happened at last! I congratulate you cordially on your good fortune and your decision, and wish you all the bliss a girl dreams about. And may my card readings not come true, needless to say; rather, you shall become a nimble, capable little housewife, for *his* happiness and as an example to all the world. One of these days, if God does not will it differently, I'll peep into your little nest and with the conscientiousness of an old aunt I'll check out everything to my satisfaction and I'll let my critical eyes dart in all directions.

My Dockerl arrived here yesterday together with her sister, so that I hang out at her place all day long, as always. Neither of us two has gotten a job and we support ourselves by private lessons -- when we can pick up some, which is still very questionable.

Is this not a journeyman's or even a gypsy's life? But I believe that we'll remain cheerful in it as ever.

I hope that you will both come again to Zurich, so that the old friendship does not become rusty. Also, I would very much like to see what sort of figure you cut as "the better half." How you will be envied by your spinster college friends!

Once again, cordial greetings and congratulations from your

Albert Einstein

82. QUESTIONNAIRE FOR MUNICIPAL CITIZENSHIP APPLICANTS

[11-26 October 1900]

CITY OF ZURICH. (MUNICIPAL COUNCIL)
QUESTIONNAIRE FOR CITIZENSHIP APPLICANTS.

1) Last and first name? *Albert Einstein*
2) Present residence ? *Zurich Dolderstrasse 17*
(city, street, street number)

3) Previous residences in Switzerland and duration of stay in them (from when to when)? *I lived in Aarau October 1895 - October 1896*
4) Religion? *None*
5) Profession for which trained? *Teacher of mathematics and physics*
6) Present occupation? *I am giving private lessons in mathematics until I obtain a permanent position.*
7) Are you self-employed or an employee?
8) By whom have you been employed since coming to Switzerland? (for each case from when to when)? *Oct. 1895 - Oct. 1896 I attended the Kantonsschule in Aarau and graduated from there. Until Summer 1900 attended and graduated from the Polytechnikum in Zurich.*
9) Are you insured? *No.*
   a. against death?
   b. against accident?
   c. against illness?
   (where, for what amount, and since when?)
10) Who is your family physician? *I have not yet had an opportunity to consult one.*
11) Have you served in the military? *No*
12) Have you served in the fire brigade? *No*

[ ]

Albert Einstein

Doct. thesis under Prof. Weber

Does now his doctoral thesis under Prof. Weber, who undoubtedly can provide the best information in this matter.

F. Bodmer

83. MILEVA MARIĆ TO HELENE SAVIĆ, WITH A POSTSCRIPT BY EINSTEIN

Zurich, 11 December 1900

My dear Helene!

Don't wonder why I did not answer your kind little letter for such a long time even now I do it with a heavy heart. Albert is leaving one of these days and is taking half of my life with him. It is better so for his career and I can't stand in its way, I love him too much for that, but only I know how much I suffer because of it. Both of us have had much to endure lately, but the forthcoming separation is almost killing me. I'll have a lot to tell you once we see each other, now I don't want to let filth spoil your most beautiful days.

I rejoice with you and am happy that you are so happy and that you have found that which presents you life in the most marvelous light. When are we going to reach the point at which we'll be allowed to acknowledge our love before the whole world, it almost seems to me that I'll not live long enough to see it. [...]

Miza

[...]

A friendly greeting to you and your husband from your
Albert Einstein

[...]

"CONCLUSIONS DRAWN FROM THE PHENOMENA OF CAPILLARITY"

13 December 1900

(Text in Vol. 2)

84. MINUTES OF THE MUNICIPAL NATURALIZATION COMMISSION OF ZURICH

27th session Friday, 14 December 1900

Present: Messrs Bodmer-Weber, President, Weber, Wirz, Kern, Müller (Lehrer), Sidler, Kuhn, Grether.
Excused absent Mr. Nägeli.
[...].
Appeared and testified by summons:
[...]

4. ALBERT EINSTEIN, reviewer Mr. Weber

I have no property, but I have a small income from about 8 private lessons a week. As an auditor at the Polytechnicum I carry accident insurance. I am in possession of expired documents as I emigrated at the age of 15. Here I have a residence permit, and Mr. G. Maier and one Mr. Bernheim deposited a personal bond. My parents have been living in Milan for the last 6½ years. I live at Dolderstrasse 17 at Mrs. Hägi's and board with a family. Am teetotaler. For the naturalization fee, a savings book of the Kantonalbank showing a deposit of 800 Fr. is submitted.

It is resolved: 1. To recommend the request to the Municipal Council.
2. Oral notification

[...]

85. MILEVA MARIĆ TO HELENE SAVIĆ

Zurich 20.XII.1900

My dearest little Helene!

[...]

As you will see at the end of the letter, Albert is still here and is going to stay here until he finishes his doctoral thesis, which will probably take until Easter, and only then is he to become a grass-widower. Of course, it is very hard for me that we must separate, but if things take their natural course I'll bear it with courage. But what utterly depressed me was the fact that our separation had to come about in such an unnatural way, on account of slanders and intrigues and all kind of things. Oh, let me tell you, even if I were to write all day long, I wouldn't be able to describe to you how much I suffered during that time, and Albert no less. But now everything is back on track, thank God. Albert's parents were a little bit behind it, you can imagine how bad it felt to be attacked from that side. But I don't want to waste any more words on that, on the ugly abominable world. I am happy that he loves me so much, and what else do I need. He is going home for Christmas, which really frightens me, but it just cannot be otherwise.

A few days ago I received a photograph of you and your husband from your little sister Alma. Thank you very much for it, I was yearning for it for a long time. We find that both of you are excellently depicted, I was especially happy that your dear eyes look so happily and contentedly into the world.

Albert has written a paper in physics that will probably be published very soon in the physics Annalen. You can imagine how proud I am of my darling. This is not just an everyday paper, but a very significant one, it deals with the theory of liquids. We sent also a private copy to Boltzmann, and would like to know what he thinks about it, let's hope he is going to write to us.

[...]

Since Albert would like to add a few words, I must now stop. I embrace you

Miza

and regards to your husband.

86. TO HELENE SAVIĆ

[Zurich, 20 December 1900]

Dear Mrs. Savić (+ $\frac{1}{9\frac{1}{2}}$, I hope)!

I'll not vent on you my impotent resentment over not getting to read Dockerl's letter to you; I'll rather rejoice a little with you in your new happiness. One can really rejoice in seeing you two in the photograph, and so we did. Your little sister has sent us the photograph, and a nice letter to go with it...but when will we see the originals? Very soon in Zurich, I hope.

But what do you say to our separation?

The lass cried
<because> the lad must split
but we thought
we won't grumble about it!

Thus we are staying together, the same as before, at any rate until Easter and also thereafter.....

To you I wish all the happiness, to your operated sister a good recovery, and to ourselves another friendly letter from you soon. Cordial greetings to you and your old folks from your

Albert

87. MILEVA MARIĆ TO HELENE SAVIĆ, WITH A POSTSCRIPT BY EINSTEIN

Zurich, Tuesday [8 January - 19 March 1901]

Dearest Helene!

[...]

But I hope that we will see each other in Zurich this summer as you wrote, i.e., provided that we will still be in Zurich. The truth is, we haven't yet the slightest idea what fate has in store for us. Albert applied for a practical job in Vienna. Since, after all, he should earn some money, he wants to improve himself in theoretical physics while holding a job so that he can become a university professor later on. We don't know, however, what will come of that. Likewise, only the gods know what will become of me, whether I'll really get a job at a girls' Gymnasium.

We still live and work as before. During these last few days we have been sledding a lot on the Zürichberg, as you see, we still have our innocent passions. Albert felt in seventh heaven whenever we went downhill "like the devil."

[...]

Miza

Dear Frau Leni [diminutive of Helene]!

Since nothing else comes to my mind and I wish to write to you something all the same, I am sending my regards to you and your old man.

Albert E...

88. REPORT OF THE SCHWEITZERISCHES INFORMATIONSBUREAU

C[ity] of Z[urich], 30.I.1901

No. 209

To: Directorate of Interior
of the Canton ZURICH

Information
on
Albert Einstein stud. math.
ZURICH V, Unionstrasse 4

We have instituted inquiries about Albert Einstein both here and in Milan. Subject designates himself as a student of mathematics from Ulm, he lives at the above-mentioned address as a lodger, and he did not offer any evidence as to his economic circumstances to the extent that we here tried to learn about them, except that it could be established that he does not yet have any income from his own work.

Regarding the report that arrived from Milan today, it follows from it that he does not have resources from the parental side either. They write:

"The Einstein family lives in Milan; father Einstein of the firm Einstein & Co. has his office in the Via Bigli No. 21 and the shop (for the manufacture of electric motors) in the Via Lecchi No. 106 and he has resided in Milan for the last 5 years; before that he was associated with a certain Garoni in Pavia, but the firm has been liquidated because of insufficient business. In Milan Einstein senior seems to do somewhat better, but he does not own any real estate and Einstein junior certainly cannot count on financial help from his father. While one can say about Einstein senior that he provides for his family adequately and properly, all his expenses must be paid out of his earnings from his work; thus, at least as far as it is known, he does not have any real property."

Herewith we conclude our report.

209

89. DEDICATION TO FRIEDRICH MÜHLBERG

[ca. March 1901]

In grateful remembrance I recall the promise you extracted from me then and am sending you my first publication.

90. TO OTTO WIENER

Zurich, 9 March 1900 [1901]

Highly esteemed Herr Professor!

Last summer I completed my studies at the mathematical-physical department of the Zurich Polytechnikum, and since I would like to expand and in some ways complete the knowledge which I acquired by attending lectures, studying the classics, and working in the physical laboratory, but am totally lacking the necessary means, I am taking the liberty of asking you whether you might need an assistant. A few days ago there appeared in Wiedemann's Annalen a short paper of mine titled "Theoretische Folgerungen aus den Kapillaritätserscheinungen" ["Theoretical conclusions drawn from the phenomena of capillarity"]

I would appreciate if you could drop me a few lines and let me know about my prospects of getting such a position now or possibly next autumn. Respectfully yours

Albert Einstein
Dolderstr. 17
Zürich-Hottingen

91. MILITARY SERVICE BOOK

Zurich, *13 March 1901*

MILITARY SERVICE BOOK

[...]

III. Health examinations.

1. Findings of the examination commission.

Div.- and recruiting district *VI/6*   Examination control No. *1309*

Body height *171.5* cm   Upper arm *28* cm

Chest circumference *87* cm   Visual acuity r $\frac{1}{3}$ [...] *1.5* [.] l $\frac{1}{2}$ [...]

Diseases or defect:

Varices [varicosis] *Pes Planus* [flat feet]

Hyperidrosis ped. [sweaty feet]

Decision of the examining commission:
Unfit A.

Secretary of the examining commission:
District Command Zurich.

[...]

92. TO WILHELM OSTWALD

Zurich, 19 March 1901

Esteemed Herr Professor!

Because your work on general chemistry inspired me to write the enclosed article, I am taking the liberty of sending you a copy of it. On this occasion permit me also to inquire whether you might have use for a mathematical physicist familiar with absolute measurements. If I permit myself to make such an inquiry, it is only because I am without means, and only a position of this kind would offer me the possibility of additional education.

Respectfully yours

Albert Einstein
Via Bigli 21
Milan
Italy

93. TO MILEVA MARIĆ

Milan. Saturday. [23 March 1901]

My dear Doxerl!

I got a sign of life from you so soon, already on the first day. This is an ugly story with Riecke; I have more or less given up the position for lost. I can hardly believe that Weber would let such a good opportunity pass without doing some mischief. Following your advice, love, I wrote to Weber so that he should at least know that what he does he cannot do behind my back. I wrote to him that I know that my appointment now depends on his report alone. I am pretty curious what Ostwald is going to write.

An original idea occurred to me on the trip. It seems to me that it is not out of the question that the latent kinetic energy of heat in solids and liquids can be conceived of as the energy of electrical resonators. In that case the specific heat and the absorption spectrum would have to be interrelated. The law of Dulong & Petit would be valid for those substances whose smallest parts show a certain total resonance in the electro-optical sense. In fact, all substances that satisfy the law of Dulong & Petit are almost totally opaque & seem to display almost the same spectrum when heated. On the other hand, the organic substances, which, as we have seen, display relatively small specific heats, are all transparent & show continuous absorption spectra, while, for example, Hg agrees quite well with Dulong - Petit's law and is completely opaque. I almost believe that the following law holds: Dul[ong]'s law is satisfied by opaque substances only. Transparent ones have always a smaller kinetic energy. Unfortunately, gases can probably not be used for solving this puzzle because of the inconstancy of phenomena in their case. However, compounds of great "internal" energy do show band-like absorption spectra. What is the story with the specific heat of glass considering its composition. It would have to have a small molecular heat, compared with its molecular number. See whether you can find something about that!

What are you up to, little imp. Brew yourself often some nice coffee and don't deny yourself anything. I think much about you (entirely unspurred) and it still makes me happy that you were so cheerful on the last day we spent together. While we are at it, let me immediately kiss your darling little trap so that it [the happiness] won't pass.

Imagine what I left in Zurich! My nightshirt & my wash things . toothbrush & comb & hairbrush. Do send everything to my sister (Töchterheim Aarau). She can later take them along with her.

It's lucky that we didn't make the Axenstrasse. This would have been a nice mess. It was not even possible to see the other shore.

I was traveling with two young fellows during my trip. Suddenly it turns out that one of them studied mathematics & physics in Göttingen for 4 semesters. You can imagine how I quizzed him about the conditions there. We also did a lot of shop-talk about the theory of cognition. He said that Riecke is a very friendly jovial gentleman & I would have very little to do as his assistant......if if....... You know the little song we so often sang together.

Here I have hardly left the house so far, and am living a quiet life instead, so that my nerves may calm down a little bit. My old folks are also contributing the best they can; the poor souls have constant aggravation and worries because of the accursed money. My dear uncle Rudolf (the rich one) picks on them terribly.

Be industrious, my love, and find yourself a friendly little room in which you feel comfortable. One kisses equally well as a little doctor and professor. Did you also send a paper to Wenger?

Affectionate greetings and kisses from

Johonzel

94. TO MILEVA MARIĆ

Milan Wednesday. [27 March 1901]

My dear Miezchen!

Thank you very much for your little letter and all the true love that's in it. I kiss and hug you for it from all my heart, exactly the way you would want it & are entitled to, love. Riecke's rejection did not surprise me, and I am also firmly convinced that Weber is to blame. Because the excuse is too implausible, and he doesn't mention a thing about the second position.

I am convinced that under these circumstances it wouldn't make any sense to write again to professors, because I am certain that all of them would turn to Weber for information once the matter had advanced far enough, and he would again give a poor reference. I'll turn to my former teachers in Aarau and Munich, but mainly I'll try to get an assistant's position in Italy. First of all, one of the main obstacles is absent here, i.e., anti-Semitism, which in German countries would be as unpleasant as it is obstructive, and, second, I have quite good connections here. That is to say, Mr. Ansbacher is a close friend of the professor of chemistry at the Milan Polytechnicum, and, further, Michele's uncle is a professor of mathematics. It is true that Michele is a terrible schlemiel but I'll grab him by the scruff of his neck and drag him along to his uncle & once I am there I'll do my own talking. At present Michele is at his parents' in Trieste with wife and child and will only be back in about 10 days. You need not fear that I will say a single word about you to him or anyone else. You are and remain to me a shrine which nobody is allowed to enter; I also know that of all the people it's you who loves me most deeply and understands me best. I can also assure you that nobody here would dare or want to say anything bad about you. How happy and proud I will be when the two of us together will have brought our work on the relative motion to a victorious conclusion! When I see other people then it really strikes me how much there is to you. On the evening of the day before yesterday, Michele's director, with whom we are rather well acquainted, was at our house for music making. He said how totally unusable and almost mentally incompetent [not responsible for his actions in a legal sense] Michele is, despite his extraordinarily extensive knowledge. The most delectable is the following little story, whose veracity can be vouched for, since the person in question knows about my friendship with Michele & has reason to fear that the thing will be reported to him... Once again, Michele had nothing to do. So his principal sends him to the Casale power station to inspect and check the newly installed lines. Our hero decides to leave in the evening, to save valuable time, of course, but unfortunately he missed the train. The next day he remembered the commission too late. On the third day he went to the train on time, but realized, to his horror, that he no longer knew what he had been requested to do; so he immediately wrote a postcard to the office, asking that they should wire him what he was supposed to do!! I think the man is not normal.

Concerning the problem of specific heat, which at the same time encompassses the relationship between temperature and radiation process, I can now think of very simple consequences for metals, which perhaps might be checked using experiments that have already been done. Let the amplitude of a train of waves progressing with a certain wave length in the direction $+x$ be $Ie^{-\alpha x}$, where I is a constant. Further, if N is the number of radiation resonators (atoms) in a unit volume, then α/N shall be independent of the nature of the substance and linearly dependent on the temperature. Hence α/N would be a function of the form $L_1(\lambda)T + L_2(\lambda)$ which is independent of the nature of the metal.

First we would have to investigate whether α can be determined by experiments on reflected light, and to what extent the experiments done so far can be used for deciding the question. I burn with desire to work my way into this, because I hope that it will be possible to make a gigantic step in the exploration of the nature of latent heat. Don't forget to look up to what extent glass obeys the law of Dulong and Petit.

Keep my umbrella for the time being. We will then see what to do about it. If only I would succeed in getting a position, so that we can go on a little trip in summer. Let's hope for the best.

Tender greetings and kisses, my dear little dumpling, from your

Albert

How are things with the new pad-to-be?

95. TO WILHELM OSTWALD

Milan, 3 April 1901

Esteemed Herr Professor!

A few weeks ago I took the liberty of sending you from Zurich a short paper which I published in Wiedemann's Annalen.

Because your judgment of it matters very much to me, and I am not sure whether I included my address in the letter, I am taking the liberty of sending you my address hereby.

Respectfully, yours truly

Albert Einstein
cand[idatus] phys[icae]
Milan
Via Bigli 21

96. TO MILEVA MARIĆ

Zurich [Milan] Thursday. [4 April 1901]

Dear Doxerl!

It's already a long time ago since I received your dear good little letter & could not yet answer it, my days are so filled up, mostly with stupid stuff. Secretly I look forward to being away from home again because it is difficuit to work here solidly.

About Max Planck's studies on radiation, misgivings of a fundamental nature have arisen in my mind, so that I am reading his article with mixed feelings. On the other hand, I have in my hands a study by Paul Drude on the electron theory, which is written to my heart's desire, even though it contains some very sloppy things. Drude is a man of genius, there is no doubt about that. He also assumes that it is mainly the negative electric nuclei without ponderable mass which determine the thermal and electric phenomena in metals, exactly as it occurred to me shortly before my departure from Zurich.

Michele arrived with wife and child from Trieste the day before yesterday. He is an awful weakling without a spark of healthy humaneness, who cannot rouse himself to any action in life or scientific creation, but an extraordinarily fine mind, whose working, though disorderly, I watch with great delight. Yesterday evening I talked shop with him with great interest for almost 4 hours. We talked about the fundamental separation of luminiferous ether and matter, the definition of absolute rest, molecular forces, surface phenomena, dissociation. He is very interested in our investigations, even though he often misses the overall picture because of petty considerations. This is inherent in the petty disposition of his being, which constantly torments him with all kinds of nervous notions. The day before yesterday he went on my behalf to see his uncle Prof. Jung, one of the most influential professors of Italy & also gave him our paper. I met the man once before & must admit that he impressed me as quite an insignificant person. He promised that he will write to the most important professors of Italy (physicists), Righi & Battelli, on my behalf, i.e., ask them whether they need an assistant. This is already quite a lot, because he seems to be on very friendly terms with them. In addition, I applied at the

Polytechnikum Stuttgart, where a position is vacant & wrote again to Ostwald. Soon I will have honored all physicists from the North Sea to the southern tip of Italy with my offer!

You were absolutely right, love, to go again to the Engelbrechts. Judging by previous experience, it is still the best place to be. If I will have earned some money by summer, we will surely take our little trip to Venice or somewhere else together. How delighted I would be! After all, I am quite a stranger here & now recognize quite clearly what the little sweetheart's love is compared with parental love. This is as different as day and night. I kiss you therefore with all my heart and you should know that your devotion makes me so happy that without it my life would be unspeakably bleak. You are right to go often to concerts, particularly to the splendid mass of Bach. Write me how did you like it.

But now I must be off to the library, otherwise it will be getting too late.

Be kissed and totally crushed by your

Johonzel

97. TO MILEVA MARIĆ

Milan Wednesday. [10 April 1901]

Dear Miezchen!

If you knew better your power over me, you little witch, you would not constantly be afraid that I might keep back all sorts of things from you, because this is really not my intention. I also want to tell you immediately, love, that my courage and my good cheer have not been broken at all, especially because I see from your letter that you are invariably cheerful. So, today I am going to give you a detailed report about myself because I see that you like that.

Last week I studied electrochemistry and chemical reactions from Michele's "Ostwald", and the electron theory of metals in the library. It's easy to explain what is setting me against Planck's considerations on the nature of radiation. Planck assumes that a completely definite kind of resonators (fixed period and damping) causes the conversion of energy to radiation, an assumption I cannot really warm up to. Maybe his newest theory is more general. I intend to have a go at it. Dude's theory of electrons is a kinetic theory of electric and thermal phenomena in metals, entirely in the spirit of the kinetic theory of gases. If it only weren't for the stupid magnetism, with which we know so little what to do! Still, I do believe that Dude is on the right track, and his conception actually receives quite creditable confirmation by experiment. I'll tell you more about it another time. I have also somewhat changed my idea about the nature of latent heat in solids, because my views on the nature of radiation have again sunk back into the sea of haziness. Perhaps the future will bring something more sensible!

Ostwald hasn't written to me (ever), neither has a professor in Stuttgart to whom I had turned, and there is just as little in prospect for me in Italy. But I am not the least dicouraged & have already weaned myself of the anger, which, after all, stemmed mainly from injured vanity. Battelli is in Pisa & Righi in Bologna. Prof. Jung, Michele's uncle, promised that he'll recommend me there. Since then I haven't learned anything. But, as I said, I am not put out, otherwise I certainly would already have poured out my heart to you, you dear, good soul, the way I got accustomed to a long time ago.

Now, I'll also tell you why I have so much to do. All the time I am performing the duties of a cicerone. For Prof. Winteler, to whom of course I must devote myself a great deal, is now here for the Easter vacation. He is an old village school principal, regardless of what he says, but intelligent all the same, and above all unprejudiced. He ignores the "casus belli," saying, girl matters..... private matters, and prefers to talk with me about other things. I must provide similar services also to a couple of ladies who are visiting with Mrs. Ansbacher. They say, correctly:"Albert has time, after all... he is also a good fellow."

Maya is now also here, and is very venomous toward me. To think that girlish unselfishness is so alien to her! In contrast, how good you are, my dear, faithful girl! Therefore, we will make our little summer trip together *for sure*, even if we have to steal the funds for it. Just keep the money I gave you, you are the best safekeeper. Also, nobody needs to know that you have something of mine.

I am very glad that you like so much to be at Miss Engelbrecht's again. She is also one of the few who deserve to be called human, she is a capable person.

Now it's again your special turn, love. Be kissed, hugged, and loved, the way your faithfulness deserves it. I think so often about it every day. Dear Miezchen is now working hard again, but in the evening I think, now she thinks of me with love, and kisses her pillow in the bed. I know how it's done!

A tender greeting from your

Albert

98. TO HEIKE KAMERLINGH ONNES

Milan, 12 April 1901

Esteemed Herr Professor!

I have learned through a friend from college that you have a vacancy for an assistant. I am taking the liberty of applying for that position. I studied at the department for mathematics and physics of the Zurich Polytechnikum for 4 years, specializing in physics. I obtained there my diploma last summer. Of course, I will make my grade transcripts available to you with pleasure.

I have the honor to submit to you by the same mail a reprint of my article that has appeared recently in Annalen der Physik.

Respectfully,

Albert Einstein

99. HERMANN EINSTEIN TO WILHELM OSTWALD

Milan, 13 April 1901

Esteemed Herr Professor!

Please forgive a father who is so bold as to turn to you,

esteemed Herr Professor, in the interest of his son.

I shall start by telling you that my son Albert is 22 years old, that he studied at the Zurich Polytechnikum for 4 years, and that he passed his diploma examinations in mathematics and physics with flying colors last summer. Since then, he has been trying unsuccessfully to obtain a position as Assistent, which would enable him to continue his education in theoretical & experimental physics. All those in position to give a judgment in the matter, praise his talents; in any case, I can assure you that he is extraordinarily studious and diligent and clings with great love to his science.

My son therefore feels profoundly unhappy with his present lack of position, and his idea that he has gone off the tracks with his career & is now out of touch gets more and more entrenched each day. In addition, he is oppressed by the thought that he is a burden on us, people of modest means.

Since it is you, highly honored Herr Professor, whom my son seems to admire and esteem more than any other scholar currently active in physics, it is you to whom I have taken the liberty of turning with the humble request to read his paper published in the Annalen für Physik and to write him, if possible, a few words of encouragement, so that he might recover his joy in living and working.

If, in addition, you could secure him an Assistent's position for now or the next autumn, my gratitude would know no bounds.

I beg you once again to forgive me for my impudence in writing to you, and I am also taking the liberty of mentioning that my son does not know anything about my unusual step.

I remain, highly esteemed Herr Professor, your devoted

Hermann Einstein

100. TO MARCEL GROSSMANN

Milan, 14 April [1901]

Dear Marcel!

When I found your letter yesterday, I was deeply moved by your devotion and compassion which did not let you forget your old luckless friend. I really believe it quite unlikely that anyone had better colleagues than I had in you and Ehrat. I don't have to tell you that I would be delighted to get such a nice sphere of activity and that I would spare no effort to live up to your recommendation. I came here to my parents three weeks ago in order to search from here for an assistant's position at a university. I could have found one long ago had it not been for Weber's underhandedness. All the same, I leave no stone unturned and do not give up my sense of humor...God created the donkey and gave him a thick hide.

We have here a splendid spring, and the whole world smiles at one so happily that one automatically sheds the old hypochondriac self. In addition, my musical acquaintances protect me here from getting sour.

As for science, I have a few splendid ideas, which now only need proper incubation. I am now convinced that my theory of atomic attraction forces can also be extended to gases, and that it will be possible to obtain the characteristic constants of almost all elements without great difficulty. That will then also bring the problem of the inner kinship between molecular forces and Newtonian action-at-a-distance forces much nearer to its solution. It is possible that experiments already done by others for other purposes will suffice for the testing of the theory. In that case I shall utilize all the already existing results in my doctoral dissertation. It is a glorious feeling to perceive the unity of a complex of phenomena which appear as completely separate entities to direct sensory observation.

Please give my kindest regards to your dear family and my warm thanks to your father for his efforts and the trust he had shown by recommending me. Friendly greetings from your

Albert Einstein
Via Bigli 21
Milan

101. TO MILEVA MARIĆ

Milan Monday. [15 April 1901]

My dear Doxerl!

Don't be angry for my not following your summons to Lugano. At the end of the last week I was in the dumps because, once again, several job hunts of mine were not showing any progress. But just wait, love, in a few weeks we shall see each other all the same - you are surprised, aren't you? Yesterday I received a letter from Prof. Rebstein from the Technicum Winterthur asking me whether I would like to substitute for him from 15 May to 15 July, because he is due for military service. You can imagine how gladly I'll do that! It's true that I have to teach about 30 hours a week, including even descriptive geometry, but the brave Swabian is not afraid. But that's not all. The evening before yesterday I received a letter from Marcel in which he informs me that probably I will soon get a permanent position in the Office for the Protection of Intellectual Property [patent office] in Bern! Isn't this almost too much at once? Imagine what a wonderful job this would be for me! I'll be mad with joy if something should become of that! Think of it, how nice it is of the Grossmanns to have exerted themselves on my behalf even now. This Rebstein is probably Hertzog's former assistant, whom we knew, actually.

As for science, I've got an extremely lucky idea, which will make it possible to apply our theory of molecular forces to gases as well. You certainly remember that the force function appears explicitly in the integrals that have to be evaluated for the calculation of diffusion, thermal conduction & viscosity. Hence, with gas molecules, *only* our constants c_α are necessary for the calculation of these coefficients for ideal gases, and one does not have to venture into the theoretically so uncertain area of deviations from the ideal gas state. I can hardly await the outcome of this investigation. If it leads to something, we will know almost as much about the molecular forces as about the gravitational forces, and only the law of the radius will still remain unknown. Unfortunately, I must also admit that this idea for the investigation of salt solutions rested on such a weak basis that I think that one should first restrict oneself to the investigation of infinitely dilute solutions, in which an interaction between the molecules of the dissolved substance does not yet occur. One can so determine a great number of c_α's, which could be used for an approximate verification of the hypothesis of the kinship with gravitation. It is possible that information about the law of action itself will more likely be provided by the quantities

$$\frac{\gamma - \frac{d\gamma}{dT}}{\text{volume}}$$

and the integrals from the theory of gases. Could you send me Kirchhoff's Heat. I would be pleased to send the *Popular Books on Natural Sciences* directly to your sister, if you don't mind. To what address should they be sent?

And how are you, dear girl? You shouldn't invest the savings for that altruistic goal, we may need all of it in summer on the Simplon. Oh, how glad I am! Now we will quite certainly be able to go. One really has to have respect for the Chinesenschn [Chinese Schn. -- unidentifiable abbreviation]. He now fills me too with respect. So, now you are my little frog! We shall see how things stand with that. Professor Winteler is leaving today, the Bessos are moving tomorrow to Trieste - though I didn't overestimate his erudition, I overestimated by far his other qualities. He is a creature without marrow and bone.

Tender kisses from your

Albert

Best regards to Miss Engelbrecht!

102. TO MILEVA MARIĆ

Zurich [Milan] Tuesday. [30 April 1901]

My dearest little child!

I don't give in! You must absolutely come to see me in Como, sweet little witch. This will take very little of your time and cause me heavenly delight. We will be back in 3 days and can arrange it so that Sunday too is still included. You'll see how lively and gay I have become and how I've got rid of all the brow-knitting. And I love you again so much! It was only because of nervousness that I was so awful to you. You will hardly recognize me, I've become so lively and gay and am longing very much to see my dear good Doxerl again. Don't you worry about the position in Zagreb, if they mess it up there. You are 1000 times more important to me than you could be to all the people of Zagreb! Who is standing in your way there, tell me a little about it. If you don't get it, that position, but I really get to be employed in Bern, then I'll appoint you herewith my dear little natural scientist. No need for you to go to a one-horse town, dear girl, I value my "old pair of boots," as you always said, more than you think. Also, you need not envy any of your girl friends, because as long as I have any desire and strength in me, I will feel happy being yours, and you will be a little shrine to me. And my happiness is your happiness. If you knew what you mean to me, you wouldn't envy any of your girl friends; because in all modesty I believe that you have more than all of them. But be it as it may, come to me to Como and bring along my blue dressing-gown, into which we can wrap ourselves & by no means forget your opera glasses. In addition, bring along a gay, light little heart and a fresh head. You never had such a marvelous trip, this I already can promise you, even if it rains cats and dogs. As soon as I get a definitive report from Winterthur, I'll write you immediately, so that you can determine the day and hour when I may wait for you.

At present I am again studying Boltzmann's theory of gases. Everything is very nice, but there is too little stress on the comparison with reality. I think, however, that O. E. Meyer has enough empirical material for our investigation. If you once go to the library, you may check it. But this can wait until my return to Switzerland. In general, I think that the book deserves to be studied more carefully.

It occurred to me recently that when light is generated, direct conversion of motional energy to light may take place because of the parallelism kinetic energy of the molecules - absolute temperature - spectrum (radiating space energy in the state of equilibrium). Who knows when a tunnel will be dug through these hard mountains! I am very curious whether our conservative molecular forces will hold good for gases as well. If here too the mathematically so unclear concept of molecular size does not manifest itself in the formation of the trajectories of molecules coming close to each other, but the molecule can be conceived as center of force. We shall get quite a precise test of our view.

Hearty kisses from your

Johonzel

103. FROM MILEVA MARIĆ

[Zurich, 2 May 1901]

My dear Johannzel!

Yesterday I sent you the assent to our trip, and I was looking forward to it so much, but don't be angry with me if I cancel it today. I got a letter from home today that robs me of all desire, not only for having a good time but for life itself. But don't let that disturb you, and take the trip alone, since you have looked forward to it for so long, we will perhaps do something together later on. I will lock myself in and work hard, because it seems that I cannot have anything else without punishment; but I also don't need anything and will get as used to this as the Gypsy [to] his horse. It doesn't matter, now good-bye, sweetheart, be cheerful and if you find lovely flowers, bring me a few. -- Greetings and kisses from your

Dockerl

104. TO ALFRED STERN

Milan, 3 May 1901

Highly Esteemed Herr Professor!

If you by chance ever happened to think of me, you surely considered me very ungrateful because I had left Zurich without even saying good-bye to you. Though this embarrassing thought often crossed my mind, I couldn't bring myself to write to you, because I always wanted to wait until I'd be able to report some pleasant news about myself -- and this has been going on exactly until today. That is to say, I have been asked to take over the teaching of mathematics at the Technikum Winterthur from 15 May to 15 July, as their professor will be on military duty during that time. I am beside myself with joy about that, because today I received the news that everything has been definitely arranged. I have not the slightest idea as to who might be the humanitarian who recommended me there, because from what I have been told, I am not in the good books of any of my former teachers, and I did not apply for the job but received an invitation. There is also a chance that I'll get a steady job in the Swiss patent office later on.

And now what should I say about all the generosity and fatherly friendliness with which you favored me whenever I had the privilege of visiting with you? I know that you no doubt know it and that you do not want to hear about it. But that much is certain, nobody has been as kind to me as you have been, and more than once I went to your place in a sad or bitter mood and always regained there my joyfulness and inner equilibrium. - But so that you do not laugh at me too much, I must add immediately that I know quite well that I am a cheerful fellow and barring an upset stomach or something of that kind, I have no talent whatsoever for melancholic moods.

I hope that your wife is quite well again, that she is so well, in fact, that you were able to take a trip during the Easter vacation, as you have frequently done before. Since Miss Dora has *certainly* passed her examination splendidly, I am sending her my cordial congratulations.

With Mrs. Ansbacher I have visited very often during my stay here, mostly while her sister from Augsburg was here with her daughter, who studies music. Luigi is now taking an additional semester in Leipzig, after having completed his studies in Pavia last year. He finds his teachers there excellent, incomparably better than those in Berlin, of course I do not know which professors he has in mind.

Shortly I will cross Spluegen on foot so as to combine the pleasant duty with a nice pleasure. When I then get to Zurich I will not deprive myself of looking you up. With many cordial greetings to you and your dear family, I remain your devoted

Albert Einstein

105. FROM MILEVA MARIĆ

[Zurich, 3 May 1901]

My dear Johannzel!

I got your d.[ear] little letter today, from which I see with astonishment that you didn't get my little note of assent. Did it indeed get lost or did something else happen to it? But I hope that in the meantime you received it after all. I also wrote you a little postcard yesterday in the worst of moods, because of a letter I got. But as I read your letter today, I became a little more cheerful, since I see how you love me, and I think that we'll make the little trip after all. <You mustn't assume>. So, I'll arrive in Como Sunday morning at 5 o'clock, because I mustn't lose an entire day on the route I already know (you are surprised, aren't you, what a good little sweetheart you have). And you will either be at the train station already, which hardly seems possible, or I'll expect you on the first train coming from Milan. Then we shall do one part of the lake on foot and botanize and chat and rejoice in each other. -- But, little sweetheart, I should know whether I am taking the same route going home, so that I can buy a round trip ticket, because it's a pity to waste the money. Why didn't you write to Winterthur once more and ask about the matter. Perhaps this is considered as self-evident there; after all, you have been asked, and have agreed. Or did they want to write to you once more?

And you love your Doxerl so much, and you long so much for her! How happy she always is with your little letters, which are full of ardent love, and which show her that you are again her dear sweetheart from before, and God! what lovely kisses she has saved for you!

How I look forward to Sunday! there are now only 2 days left until then, so don't oversleep. Awaiting you with thousand joys your suffering

Toxerline

106. TO MILEVA MARIĆ

Winterthur, Thursday evening [9 May 1901]

Dear Miezchen!

My first greeting from here shall belong to you, love. Now let me first tell you what has happened to me since we parted. First I went to the Hotel Limmathof, where there was no room for me after they scrutinized me in my dubious suit from head to toe. Then I went to the Hotel Central, where they just barely granted me shelter against advance payment of 2.50 fr. This morning Mrs. Hägi received me very amiably and helped me pack my little suitcase and wanted with all her might to force some food upon me. She is better after all than we thought. Old Stern was very happy about my having been invited here, & Maier's son, who studied mining science, is in East Siberia in a place where at a certain depth the ground remains frozen all year long & in winter the temperatures are sometimes as low as 50° below zero. - After that, I had lunch at Orsini's with one of the young men from Bahnhofstrasse & took off for Winterthur at 3 o'clock. In front of the railway station I immediately met Rebstein, who fixed a meeting for me at the Technikum for tomorrow morning at 10 o'clock, so that I can get an idea about the way I have to teach. I am looking forward to my work with great joy. Rebstein told me that he thought of me himself, and that Amberg and Ehrat recommended me to him; well-meaning people do exist, after all. Then I went to the young Wohlwend at his office, who was awfully glad to see me. I rented a room at his landlady's (Aussere Schaffhauserstr. 38) and will eat in his boarding house. You have no idea how charming and clean my room is! A large room with a double window, a verandah with a glass door and a most friendly view, parquet floor, an unspeakably comfortable sofa and beautiful carpets, a few nice pictures -- in short, really ideal. Also, everything spick and span and clean. If you could only see it.

The house is a pretty villa outside the little town, which now, in the most beautiful season, gives the impression of a flower garden.

Have you now given a good rest to your dear little feet and are you now all fresh in body and spirits again? If only I could give you some of my happiness, so that you would never be sad and wistful. I don't know yet whether I'll come Sunday, because this happens to be the only day on which I can find Grossmann at home. But I might come to you early in the morning, have lunch with you and then go to Thalweil only in the afternoon. I must think about it some more.

Now I have to fetch Wohlwend. Ardent greetings and kisses from your

Albert

107. TO MILEVA MARIĆ

[Winterthur, second half of May? 1901]

Dear Doxerl!

I bet you are puzzled by the funny piece of paper on which I am writing to you. But I think, my Doxerl doesn't mind, seeing that I don't have any other one. Don't be angry for my not having written you for so long, I simply haven't much to say that you don't know already. So I help myself out with what always stays lovely and nice. I am fond of you, my dear girl, and am looking forward to seeing you again on Sunday. We shall again spend an enchanting cozy day together. Only the thought of you gives my life here a true meaning. If only the thought had a little life and flesh and blood! How delightful it was the last time, when I was allowed to press your dear little person to me the way nature created it, let me tenderly kiss you for that, you dear good soul! How is your work going, dear sweetheart? Everything proceeding jolly well? Does old Weber behave decently or does he again have "critical theorems"? The local Prof. Weber is very nice to me and shows interest in my investigations. I gave him our paper. If only we would soon have the good fortune to continue pursuing this lovely path together. But destiny seems to bear some grudge against the two of us. But this will make things all the more beautiful later on, when all obstacles and worries have been overcome.

My parents seem to be down and out once again, because they asked me to send Maja <100> 50 fr. On 8 August they will have their silver anniversary. How sad I will feel during this little celebration! Papa again reminded Michele that he should write to me, thus far in vain!

But all that doesn't matter. After all, I have you and your love!

Thousand kisses, and an extra sweet one, from your

Albert

108. FROM MILEVA MARIĆ

[Zurich, second half of May? 1901]

Dearest sweetheart!

I have now received your second little letter as well, and am very happy, immeasurably so. How sweet you are, oh how I will kiss you, I can hardly wait for the end of the week when you are coming. I think I'll pray to little Peter to breathe a good idea into Mr. Besso. - If you were to come on Saturday, you could probably sleep at our place, because a woman is going away on Friday, I'll ask Miss Eng[e]l[brecht] she'll do it for me if it's possible. In the meantime I'll be very diligent so that I can then freely rejoice with you - my God, how beautiful will the world look to me when I'll be your little woman you'll see, there will be no happier woman in the whole world and then the man must also be as happy.

Good-bye my sweet little darling and at the end of the week come cheerfully to your

little woman

109. MILEVA MARIĆ TO HELENE SAVIĆ

[Zurich, second half of May? 1901]

My dearest Helene!

[...]

Albert has been in Winterthur since the beginning of May. On 5 May I went to Como, where a certain person waited for me with open arms and a "pounding heart." I should tell you a little bit about our trip, because it was so beautiful that it made me forget all my sorrows. We stayed in Como half a day and then proceeded by ship toward Colico. We made a little stop in Cadenabbia and visited Villa Carlotta. I have no words to describe the splendor we found there. You know, they have there a few things by Canova; and then, the splendid garden, which I especially preserved in my heart, the more so because we were not allowed to swipe even one single flower. It was the most beautiful spring when we were there, and we did not even suspect that fate had ordained that we should be riding on a sledge through snow flurries the very next day! The Splügen, which we wanted to cross, lay deep in snow, which in some spots reached a height of 6 m[eters]. Therefore we rented a very small sledge, the kind they are using there, which has just enough room for 2 people in love with each other, and the coachman stands on a little plank in the rear and prattles all the time and calls you "signora," -- could you think of anything more beautiful? We had to travel a few hours, but only up to the pass, because we wanted to dare it on foot from there. It was snowing so gaily all the time, and we were driving now through long galleries, now on the open road, where there was nothing but snow and more snow as far as the eye could see, so that this cold, white infinity gave me the shivers and I held my sweetheart firmly in my arms under the coats and shawls with which we were covered. - The descent from the Splügen was also beautiful, we had to tramp hard through the snow, but we had so much fun that we did not feel it burdensome at all. In suitable places we produced avalanches so as to properly scare the world below us. -- You too know the Rhein valley and the Via Mala, I think, they were simply magnificent, even though the weather was a little gloomy, but this did not spoil our good mood. How happy I was again to have my darling for myself a little, especially because I saw that he was equally happy! He now comes to see me each Sunday and each time we think of our dearest friend. Albert would so much like to see you and find out how the "hatching" becomes you. He was very touched when I showed him your last letter, and he said: we too want to be as happy as they are. How is your health? Take good care of yourself, the little dears like that. And your husband? Has he completely recovered? - Albert now again often suffers from his famous ailment.

[...]

I had a few spats with Weber, but we have already gotten used to it. Albert is very satisfied in Winterthur, if only he could get such a work for good. He will now try and see whether he can get a job with an insurance company through a good acquaintance of his.

[...]

Miza

[...]

110. TO MILEVA MARIĆ

Winterthur Thursday [second half of May? 1901]

My dear Doxerl!

I don't want to go to bed without answering your dear little letter, which I saw today lying on t[he] table when I came home from the school -- a most darling little letter. I am so much looking forward to Sunday with my dear Doxerl. Be cheerful, dear Doxerl, and don't worry -- you are my dear, good sweet[heart], whatever may happen.

I am not very satisfied with my theory of thermoelectricity. I am n[ot] going to publish it for the time being. Perhaps I'll write a private letter to Drude to draw his attention to his mistakes. This evening I sat 2 hours at the window and thought about how the law of interaction of molecular forces could be determined. I've got a very good idea. I'll tell you about it on Sunday.

I have not received any reply whatsoever from my sister. She is now in the midst of that awkward age of a girl's adolescence. If only she will come out of it alright. Michele hasn't yet written me either. I think I'll turn to his father and ask him whether he can find me a position in insurance. That's a stupid matter about being a starveling. [B]ut otherwise he is a splendid fellow, your sweetheart is, even if somewhat unlucky.

Writing is stupid. Sunday I am going to kiss you orally. Greetings and hugs from your

Albert

To a happy reunion! Love!

111. TO MILEVA MARIĆ

Winterthur Tuesday [28? May 1901]

My dear Miezchen!

I have just read a marvelous paper by Lenard on the production of cathode rays by ultraviolet light. Under the influence of this beautiful piece of work, I am filled with such happiness and such joy that you absolutely must share in some of it. Just be of good cheer, love, and don't fret. After all, I am not leaving you and I'll bring everything to a happy conclusion. It's just that one has to be patient! You'll see that one doesn't rest badly in my arms, even if it starts a little stupidly. How are you, love? How is the boy? Imagine how lovely it will be when we will again be able to work together totally undisturbed, and nobody will any longer be allowed to interfere! You will be amply compensated for your present worries by a lot of joy, and the days will peacefully pass by, undisturbed and unhurriedly.

I was alone all day long yesterday because Wohlwend was in Lenzburg, and I studied Wiedemann's Annalen after having taken a very lovely walk through the woods in the morning. I found there a numerical confirmation which a Dutchman had found for the fundamental principles of the electron theory, which filled me with real delight and completely convinced me about the electron theory.

Wohlwend also went to see the Wintelers, but of course said nothing there about me, only a little something to my sister, to whom I am now going to write. Distance seems to have softened her grudges against me very much. I'll invite her here for a Sunday.

How are our little son and your doctoral thesis? If I am not mistaken, Weber also once did theoretical work on the motion of heat in metal cylinders. See whether you couldn't somehow use the tables on this basis, even if only ostensibly. I think that he is cited in Heine.

Unfortunately, no one here at the Technikum is up to date in modern physics & I have already tapped all of them without success. Would I too become so lazy intellectually if I were ever doing well? I don't think so, but the danger seems to be great indeed. Unfortunately, I learned today that there is quite some competition between mathematicians in Switzerland too. In Germany it is supposed to be much worse. I have already wondered whether old Besso couldn't find a job for me in insurance. After all, he is the general manager of a company. Don't get any silly ideas, I'll do everything I can to stand up for you, love.

So, chin up, and do write soon a dear little letter to your

Johonzel

112. TO MILEVA MARIĆ

Tuesday [Winterthur, 4? June 1901]

Dear Doxerl!

What do you think is lying on the table in front of me? A long letter addressed to Drude with two objections to his electron theory. He will hardly have anything sensible to refute me with, because the things are very simple. I am terribly curious whether and what he is going to reply. Of course, I also let him know that I don't have a job, that goes without saying. I already told you what it is all about. I got a postcard from my sister. She is not coming to visit me. Imagine, the Wintelers railed against me at the Wohlwends' & said that I have been leading a life of debauchery in Zurich. -- Nothing beats the "eternal feminine." Supposedly Byland didn't behave too nicely either. He would seem to fit much better the words bellowed during the instruction hour by a good German sergeant about Napoleon I: "He was a very good soul....but stupid, stupid, terribly stupid.["]

How are you, dear sweetheart? Let me hear from you soon! Do you still remember how clumsy I was the last time? But I didn't write anything about that to good old Drude, would you believe it? How are your studies and the child and the <child> mood? I hope that all three are fine, as it should be. I am sending you special kisses so that there should never be a lack of good mood. The future will bring with it whatever the present leaves to be desired, and lots of it. If Michele doesn't write soon, I'll write him again to ask for a position for me from his strict Herr Papa. If one doesn't do splendidly, one's good friends tend to leave one in the lurch. But this is how it goes.

I really do have your little jacket. I'll bring it with me next time.

Yesterday I again played music at the place of the older miss. It was great. If only you could have been there too! You badly need a nice change. I still have to give a private lesson in algebra this evening.

I am looking forward to next Sunday. If only we could be together carefree and in good cheer for once, without any pressure on us. I believe that you cannot imagine yourself in such a situation any more than I can, you good poor girl. Affectionate kisses from your

Albert

113. TO THE DIRECTOR'S OFFICE, TECHNIKUM BURGDORF

Winterthur, 3 July [1901]

To the Director's Office!

I have learned that your institution has a vacancy for the chair of Strength of Materials and am taking the liberty of applying for that post.

I have been living in Switzerland for almost 6 years and have obtained Swiss (Zurich) citizenship during that time. In the Fall of 1896 I graduated from the Kantonsschule in Aarau and after that I enrolled in the School for Teachers of Mathematics at the Federal Polytechnikum. There, besides the usual mathematics and physics courses, I also took courses in technical subjects, such as Strength of Materials with Prof. Hertzog, and Electrical Engineering with Prof. Weber. In the summer of the past year I obtained there my specialized teacher's diploma.

Since then I have been working on investigations in the physics laboratory and on studies in theoretical physics. I also published a paper on capillarity in Wiedemann's Annalen.

Since 15 May I have been teaching mathematics at the Technikum here in Winterthur as a substitute for Dr. Rebstein, who will be absent until 15 July because of military service.

Needless to say, my records are at your disposal. For further information please contact Prof. Lüdin, Prof. Weber, Prof. Rebstein at the Winterthur Technikum, Prof. Hertzog in Zurich, and the professors of the cantonal school in Aarau. With one of the latter, Prof. Winteler, formerly of Burgdorf, I am on very friendly terms.
Respectfully

Albert Einstein
Äussere Schaffhauserstrasse 38

114. TO MILEVA MARIĆ

[Winterthur] Sunday evening [7? July 1901]

My dear Doxerl!

I have just come home from Lenzburg & found this letter from Drude, which is such an irrefutable evidence of its writer's wretchedness that no comment by me is necessary. From now on I'll not turn any longer to this kind of person but will rather attack them mercilessly via journals, as they deserve. It is no wonder that little by little one becomes a misanthrope.

But now, rejoice in the irrevocable decision that I have made! I decided the following about our future: I will look *immediately* for a position, no matter how humble. My scientific goals and my personal vanity will not prevent me from accepting the most subordinate role. The moment I have obtained such a position I'll marry you and take you to me without writing anyone a single word before everything has been settled. And then nobody can cast a stone upon your dear head, and whoever dares to do anything against you, he'll better watch out! When your and my parents are faced with the fait accompli, they'll just have to reconcile themselves with it the best they can. And as my little wife, you can peacefully rest your little head in my lap and will not have to regret the tiniest bit the love and devotion you have bestowed upon me.

Even though our situation is very difficult, I am again quite confident since I made this decision. First thing tomorrow I will write to old Besso and go to the director of the local insurance company, who will be able to give me further advice.

Affectionate kisses from your

Albert

115. TO JOST WINTELER

Winterthur, Monday [8 July 1901]

Dear Herr Professor!

I was very happy to learn from my parents' last letter that you thought of me when you heard of a vacant teaching position in Burgdorf and that you are even ready to put in a recommending word for me there. I thank you with all my heart for your friendly offices.

Immediately after the receipt of this information (last Wednesday) I wrote to the director of the Burgdorf Technikum and applied for the position. The next day my colleagues told me that this teaching position involves not only mechanics and strength of materials, but also includes instruction in machine design, for which practical experience is essential. However, this does not say that the teaching of all these subjects by *one* teacher will also go on in the future. I have not yet received a response. I indicated that they may ask my former teachers at the Aarau cantonal school about me, and, further, that you and I are personal friends. I wrote this mainly because I thought that the gentlemen there are acquainted with you and will turn to you for information. I just don't know whether it would have been pleasant for you to give an objective judgment about me -- I would find an analogous situation somewhat awkward, one has to stick rigorously to the truth, and at the same time one does not like to say anything unfavorable. But in this way you can easily refrain from giving an opinion if this seems more appropriate to you.

I have been quite exceptionally pleased with my activities here. It had never occurred to me that I would enjoy teaching as much as it actually proved to be the case. After having taught 5 or 6 classes in the morning, I am still quite fresh and work in the afternoon either in the library on furthering my education or at home on interesting problems. I cannot tell you how happy I would feel in such a job. I have completely given up my ambition to get a position at a university, since I see that even as it is, I have enough strength and desire left for scientific endeavor.

There is no exaggeration in what you said about the German professors. I have got to know another sad specimen of this kind -- one of the foremost physicists of Germany. To two pertinent objections which I raised against one of his theories and which demonstrate a direct defect in his conclusions, he responds by pointing out that another (infallible) colleague of his shares his opinion. I'll soon make it hot for the man with a masterly publication. Authority gone to one's head is the greatest enemy of truth.

But I do not want to bore you any longer with my talk. Thanking you sincerely, I remain your

Albert Einstein
Schaffhauserstr. 38, Winterthur

116. FROM MILEVA MARIĆ

[Zurich, ca. 8 July 1901]

So, you'll immediately look for a position sweetheart and take me to you! How happy I was when I read your litle letter, and how happy I still am and always will be. And if I don't infect you too, sweetheart, off with my head!. But of course, dear, it shouldn't be the worst possible position, this would make me feel awful, I wouldn't be able to stand it. Will then our assorted old folks be surprised. By the way, my sister has written to me that I should invite you to visit us during the vacation, my parents are now probably in a better mood. Wouldn't you like to come along, it would make me happy! And think of the beautiful journey we would make together! Here and there we would get off the train and go on foot a little or stay over for a short time. And then in our parts everything would be new to you. And when my parents see the two of us in the flesh in front of them, all their misgivings will evaporate.

Was it nice in Lenzburg; there was such a terrible thunderstorm on Sunday, I was afraid that you might be still underway. I hope you were already indoors, sweetheart. Actually, I also wanted to give you some cherries, but you couldn't have taken them to L[enzburg]. They are now waiting for you under lock and key. -- I am very diligent, I must now study Weber hard, and in between I am joyfully waiting for Sunday, when I can see you again and kiss you in the flesh and not only in my thoughts, and almost the way my heart commands it and everywhere everywhere.

What are you up to, sweetheart? Do you too have such awful weather.

117. TO THE DEPARTMENT OF EDUCATION, CANTON OF BERN

Winterthur, 13 July 1901

To the Department of Education!

Intending to register for the vacant chair of mechanics and strength of materials at the Burgdorf Technikum, I mistakenly applied to the director of the Burgdorf Technikum. Please let me know whether my application will be taken into account or whether I have still time to send a new one to the Department of Education. Respectfully

Albert Einstein
Schaffhauserstr. 38
Winterthur

118. FROM THE DEPARTMENT OF INTERNAL AFFAIRS, CANTON OF BERN

[Bern] Dispatched on 16 July 1901

Reply. - Your application for the vacant position of Teacher at the Burgdorf Technikum has been received by us and will be submitted together with all the other ones to the supervisory committee of the institute for evaluation.

1 one-franc bill returned.

Ritschard

119. TO MILEVA MARIĆ

Monday [Mettmenstetten, 22? July 1901]

My dear little sweetheart!

Thank you with all my heart for your dear little postcard. But Herr. Adjunkt has come to naught. I asked Haller by phone & got a negative answer. I had expected that from the start, because this is an administrative position.

I am now getting along somewhat better with my mother. But she suspects something. That is to say, she believes I'll marry you as soon as I have a job: as a matter of fact, my sister told her that that little postcard was from you.

And now, I wish you the best of luck for your exam & may it pass quickly, dear sweetheart. I think that we should then immediately take a little trip as a redress, but only over the Klausen, because we must economize a little & the cost difference is quite significant. Don't you agree? I am working on a theory of the liquid surface all the time, but totaly unsuccessfully. All my endeavors since the paper have their deficiencies. What I said in the lecture I gave you is all wrong. Some day I'll prove it to you.

It's wonderful here in Paradies. From the front verandah one has an all-dominating view which reveals new charms every time.

Unfortunately, all sorts of people (e.g., from Genoa) will again visit us here, which I really abhor. If only I were soon to get a position & we could vegetate together. This is my greatest desire.

But now my mother is here for a cup of coffee.

Affectionate kisses from your

Albert

Good luck!

120. FROM THE DEPARTMENT OF INTERNAL AFFAIRS, CANTON OF BERN

Bern, 31 July 1901

The Department of Internal Affairs to

Esteemed Sir,

While thanking you very kindly for offering your services in filling the vacant position of Senior Teacher at the Mechanical - Technical Department of the Cantonal Technikum in Burgdorf, we must inform you that you have not been elected. Your documents are returned in the enclosure.

Respectfully,

Head of the Department of Internal Affairs

Enclosures

121. FROM MILEVA MARIĆ

[Zurich] Wednesday [31? July 1901]

My dearest little sweetheart!

I've just received your dear little letter, which shows me that you are not exactly cheerful either. We are some little couple, and on top of it the people here envy us all the time, this is really the limit! Did you have, then, an open fight with your mother? Dear little sweetheart, how much you have to endure for me! And the only thing I have to give you for all that is the little bit of love that dwells in the human heart. But you know, this is not so awfully little after all, and it will compensate you for quite a few things, if this is humanly possible. And I simply cannot believe that your mother will never become reconciled with you. Her whole relation to you would then have to consist of ambition and self-love alone, and be completely devoid of any love, and such mothers do not really exist. Also, you should always remember that they still only know me as a misconception, and that it is still up to me to meet them in a light that is more pleasing to them. I think that a reconciliation will require much time and goodwill, but I am sure that it will come about. You know, I have even devised several methods we could use to start the process. E.g., if I could ingratiate myself with an acquaintance of theirs to whom they look up a little, they would essentially be conquered (this is what I assert). And I have also devised some other little things.

Do write soon to my old man, sweetheart, since I would like to leave on Saturday and they should have a little letter before I arrive home. Will you send me the letter so that I can see what you have written? - I'll be travelling with my friend Miss Buček. She also doesn't suspect the mixed feelings with which I am going on this trip. Write to my Papa just briefly, I shall then gradually break the necessary news, the disagreable ones included.

But if you think that Michele takes such an attitude toward the matter, then it might possibly be better if you do not tell him anything at all. It's simply that all of them have left behind the stage of pure humane sentiment, mired in life's banality. You see how your sister has a different attitude toward such things, even if she vacillates from time to time. I am really glad that she isn't up in arms against me like the others, but just the opposite.

I'll send the money immediately to Milan, should I send you the other stuff too? Or if we meet, once more, I'll bring it to you. There is a train that passes Mettmenstatten at 7: 56 A.M. toward Zug, stops there for a quarter of an hour, and then turns back again. Would you like to take this journey with me, sweetheart? Oh, if only I could have you once more, just exactly to my heart's desire, my dear sweet love! If you knew how I love you, you are my little all. And now goodbye and don't let it get you down too much, my darling. Think a little of your little one, and be kissed and hugged by your

D[oxerl]

Has Prof. Winteler already recommended you in Frauenfeld? Would it not be appropriate for you to introduce yourself to the pertinent people there? That's a custom with us, I don't know how it's here.

122. TO MARCEL GROSSMANN

Winterthur, Friday [6? September 1901]

Dear Marcel!

With great joy have I just read in the newspaper that you have become professor at the cantonal school in Frauenfeld. I congratulate you cordially on this success, which offers you nice work and a secure future. I too applied for this position, but, in fact, I did it only so that I wouldn't have to tell myself that I was to faint-hearted to apply; for I was strongly convinced that I have no prospects of getting this or another similar post. However, I too am now in the happy position of having gotten rid of the perpetual worry about my livelihood for at least one year. That is to say that as of 15 September I will be employed as a tutor by a teacher of mathematics, a certain Dr. J. Nüesch, in Schaffhausen, where I'll have to prepare a young Englishman for the Matura [high-school graduation] examinations. You can imagine how happy I am, even though such a position is not ideal for an independent nature. Still, I believe that it will leave me some time for my favorite studies so that at least I shall not become rusty.

Lately I have been engrossed in Boltzmann's works on the kinetic theory of gases and these last few days I wrote a short paper myself that provides the keystone in the chain of proofs that he had started. However, it is too specialized to be of interest to you. In any case, I'll probably publish it in the Annalen. On what stuff do you spend your free time these days? Have you too already looked at Schopenhauer's Aphorisms on the Wisdom of Life? This is a part of Parerga & Paralipomena, and I liked it very much.

A considerably simpler method of investigating the relative motion of matter with respect to luminiferous ether that is based on ordinary interference experiments has just sprung to my mind. If only, for once, relentless Fate gave me the necessary time and peace! When we see each other I'll report to you about it.

Give my best regards to your family and accept once more my heartfelt congratulations. Your

Albert Einstein

123. FROM MILEVA MARIĆ

[Stein am Rhein, early November 1901]

Dear little sweetheart!

I am writing you now only a few words because I am angry with the cruel fate which ordained that tomorrow I must sit alone! I am so glad that Kleiner was nice to you! And during which vacation could you maybe carry out the investigation? If you knew how glad I am, your last little letter made me very depressed. But, please, sweetheart, don't tell your sister that I am *here*. I know that she won't do any mischief on purpose, but I am terribly scared that something could happen again, as it always did. This you really could do for me, couldn't you, sweetheart, but in all seriousness! Promise this to me. Otherwise, give her my kindest regards and tell her that I was very pleased with her kind words that you showed me. But don't give her my address, sweetheart, because I am terribly scared.

Are the flowers still fresh? Did you put them in water? Don't write to your parents anything about me. Only no further tempests, it makes me shudder even to think of it. The present peace and quiet is so nice and beneficial.

But now I am not writing you anymore since I am angry because of tomorrow. So, agreed, sweetheart, keep your (trap) mouth shut about my whereabouts, tell them I am in Germany. What nice books you sent me! The Visit in Detention is marvelous. Did I laugh! I've also read the one by Forel, when I finish it, I'll write to you about it. Have you already read the one by Planck? It seems to be interesting.

But now, I kiss and hug you frightfully. Your

Dock

124. FROM MILEVA MARIĆ

[Stein am Rhein] Wednesday [13 November 1901]

My dear, wicked little sweetheart!
So now again you are not coming tomorrow! And you don't even say: I am coming on Sunday instead! But you'll surprise me then for sure, won't you? Let me tell you, if you are not coming at all, then I'll run away all of a sudden! If you knew how homesick I am, you would surely come.

But you really don't have any money left? Very nice! the man earns 150 fr., has bed and board, and doesn't have a centime at the end of the month! What would anybody say about that! But this won't serve as an excuse for Sunday, will it, and if you don't get any money by then, I'll send you some.

Is your cousin still staying with you? Has he found his ticket? Did he come especially to visit you, otherwise one doesn't travel via Schaffhausen, and has such a mishap.

You know, there was a fair in Schaffhausen yesterday, but unfortunately I heard about it too late, otherwise I would have gone there and bought something nice for you, and would have looked up your tower and, if possible, my dear sweetheart. If you knew how I would like to see him again! I think of him all day long, and in the evening all the more; and then he tells me all sorts of nice things, and then I think of him even more.

I am very curious what Kleiner will say about the two papers. He'd better pull himself together and say something reasonable. It would make me very happy if you could soon do the other one as well. I'll write to Helene. She's surely got her "tiny one" by now. I haven't written her for a long time because I couldn't bring myself to do so in those terrible times. I once wrote her a long letter and poured out all my woes; but then I tore it up, and now I am glad that I did so. I don't think we should say anything about Lieserl yet; but you too should write her a few words now and then, we must now treat her very nicely, she'll have to help us in something important, after all, but mainly because she is so nice and kind and because it will give her such pleasure. Agreed, sweetheart. - I finished reading the book by Forel. Stadler said that hypnosis is an immoral thing, and when I read the book I had exactly the same, disgusted feeling. Suggestion plays an important role everywhere, and I think that a physician is even obliged to apply it, up to a certain point. But such a violent surprise attack upon human consciousness! In my opinion, Forel differs from a quack only in that he faces his patient with more self-confidence, read impertinence, because of his more extensive knowledge. But people are such a stupid bunch. The hypnotic sleep I cannot understand, perhaps it cannot be understood at all, if it exists at all! I think that this is also suggestion, or, at best, auto-suggestion, because I consider most of the experiments he cites as dishonest (I am very sorry!). One of these days I'll tell you why.

But now farewell, my little one, my darling; do you think of me sometimes, but kindly and nicely? But you are coming on Sunday, sweetheart, aren't you, I've already stored up such a lot of kisses, if the cup runs over, all will go away. And now, hearty greetings and kisses from your

Doxerl

who is for the present right angry with you

I'll tell you a funny story, something that once happened to you!

125. MILEVA MARIĆ TO HELENE SAVIĆ

Neusatz, Kissacsergassse 20 [ca. 23 November - mid-December 1901]

My dear good little Helene!

[...]

I have even written to you several times, but my letters were so full of bitterness and ill humor that I tore them up immediately; I could not bring myself to send you gloomy news, and only such ones were at my disposal. All the same, do not get scared, dear, I am still alive and even quite merry again, and so is my sweetheart. The whole misery was due to the charming behavior of my dear mother-in-law <!> . That lady seems to have made it her life's goal to embitter as much as possible not only my life but also that of her son. Oh, Helene, I wouldn't have thought it possible that there could exist such heartless and outright wicked people! Without further ado they found it in their heart to write a letter to my parents in which they reviled me to such an extent that this was really a shame. You could probably imagine in what a conflict this resulted. In any case, it caused much suffering to both Albert and myself. Now the air has cleared to some extent, i.e., Albert's parents are not so terribly angry with him anymore. In addition we have the misfortune that Albert has not got a position; he is now in Schaffhausen, where he is employed as a tutor. You can imagine that he does not feel good in such a state of dependency. Yet, it is not likely that he will soon get a secure position; you know that my sweetheart has a very wicked tongue and is a Jew into the bargain. From all this you can see that we are a sorry little couple. And yet, when we are together, we are as merry as hardly anybody. When I was now in Switzerland, we saw each other a few times. And you know: in spite of all the bad things, I cannot help but love him very much, quite frightfully much, especially when I see that he loves me just as much.

[...]

Albert has written a magnificent study, which he submitted as his dissertation. He will probably get his doctorate in a few months. I read it with great joy and real admiration for my little sweetheart who has such a good head on his shoulders. I'll send you a copy when it gets printed. It deals with the investigation of the molecular forces in gases using various known phenomena. He is really a splendid

fellow; if only we could come together for once. But without friends it is difficult even for such a man to find any kind of secure job. Pray for us, dear little Helene, that things no longer go so terribly wrong for us!
[...]

Miza

126. TO MILEVA MARIĆ

[Schaffhausen] Thursday [28 November 1901]

Dear sweetheart!

Three days have passed without my having received a letter, and as many nights. But I am so firmly convinced that you wouldn't let me wait so long, that I definitely believe that the letter got lost. Did you receive the 2 or 3 letters of mine that I mailed to Katy [Hungarian spelling for Kać] and the one I mailed to Neusatz [German name for Novi Sad]? I almost believe that your mailmen use the letters for kindling or even......hor[r]ibile dictu [horrible to say], but I am not saying it. From now on I am going to write you in each letter that I'll write you often, so that you won't worry and will know that everything is alright except for the negligent postal service.

Except for the lack of news, I am really fine & I am almost always in good cheer. If only I could be sure that this little letter is certain to end in your hands at last. Let them occasionally check in Katy whether the letters really aren't there! I can hardly believe it.

It is very cozy in the new room, even though its only ornaments are myself and the dear red lampshade, of which Mrs. Baumer said that she wouldn't do such a tremendous job even for her Karl. But I thought to myself. My dear sweetheart would do quite different things for me as well, but so would I for her. All this I have already written to you, but who knows whether you received it. I am unspeakably happy that your parents are now somewhat calmer and have more trust in me. But I also know that I deserve it and that their Miezel will get a good husband as soon as this becomes feasible. The position in Bern has not yet been advertised so that I am really giving up hope for it.

Write me in detail how you are spending your day so that I can follow you a little in my imagination, I don't think that this should be too difficult to imagine. I am living here as if I were completely alone, since I do not see anybody privately. Almost every day I take a little walk to refresh myself, the rest of the time I spend studying Voigt's theoretical physics, from which book I've already learned quite a lot. The evening of the day before yesterday the local music teachers organized an evening of chamber music, which was delightful beyond my expectations. So far I have no report from Kleiner. I don't think he would dare to reject my dissertation, but otherwise, in my opinion, there is nothing that can be done with that short-sighted man. If I had to be at his beck and call to become a university professor -- I wouldn't want to change jobs, but would rather remain a poor private tutor.

Should I send you a book, dear sweetheart, or do you have some other fulfillable wish? Write it to me without hesitation -- but write me much. Each letter makes me very happy. They are the only human pleasure that gladdens my mind. They must substitute for wifey, parents, friends and company, and they can do it, too. Of course, it would be nicer if I could have you with me as in the good old student days in Zurich. As soon as I become a doctor, I'll put in an advertisement in order to seek a secure position. One of these days we shall have external luck as well. When you'll have your fill of being at home, come to me, you'll always be received with open arms, we will manage somehow.

Write me pretty soon and right frankly. Hearty kisses from your

Albert Johonzel

127. TO MILEVA MARIĆ

Schaffhausen Thursday [12 December 1901]

My dear sweetheart!

I received your dear belly-ache letter, which you were so nice to write me in bed. But I am not worried at all, because I see from your good mood that the problem is not so serious. Take good care of yourself and be cheerful and rejoice in our dear Lieserl, whom I in absolute secrecy, to be sure (so that Doxerl wouldn't notice it), prefer to think of as Hanserl.

There is a lot of news in my case, some good and some not. First of all, it should be noted that Kleiner hasn't written yet. The second is that Louis' mother whines terribly about our emigration project. She declared that she doesn't mind the money (she must have almost as much of it as the two of us together, don't you agree?). Also, he shouldn't cause her worries and excitement, she really has enough of those (her husband suddenly became insane in August). I advised him therefore to give up the plan. Besides, there was the danger present that I might fall between two stools if I persisted with the plan. Imagine in what an awful fix we would have been then! I therefore decided to set up myself here as comfortably as possible. I therefore went to N[üesch] and told him to give me the money for the meals, so that eventually I might save a little money. He said, aflush with rage, that he has to think about it. Then he conferred with his fine little spouse. When I returned in the evening, he was very snotty and told me, with an authoritative expression: "You know our conditions, there is no reason to deviate from them. You can be quite satisfied with the treatment you are getting." To which I said: "Very well, as you like, I have to give in for the time being. -- I'll know how to find living conditions that suit me better." (Imagine what nerve, in my position!) He understood that and softened up immediately. He noticed that I am less concerned with saving money than with not wanting to eat with him and his fine family, swallowed his rage and told me as softly as he could: "Would you be satisfied if I arranged for your meals somewhere, in a restaurant?" I immediately understood why he wanted that -- so that it shouldn't be possible to calculate how much he steals from the 4,000 fr. put out for me. So I agreed happily and took my leave, remarking that he should make the arrangements as quickly as possible, I had attained my goal, after all. They are now foaming at their mouths with rage against me, but I am now as free as the next man. I have already eaten there today, it's very cozy there, I have already found quite nice company there in the persons of two young pharmacists. Long live impudence! It's my guardian angel in this world.

Yesterday, as I happily entered the house N. for the last dinner, just before the subscription concert at which I participated, there was a letter from Marcelius on my soup plate, a very kind letter, in which he tells me that the position in Bern will be advertised within the next few weeks and that he takes it for certain that I'll get it. In 2 months' time we would then suddenly find ourselves in splendid circumstances and our struggle would be over. I am dizzy with joy when I think of it. I am even happier for you than for myself. We would be the happiest people on earth together, that's for sure. We shall remain students (horribile dictu) as long as we live, and shall not give a damn about the world. But neither shall we forget that we owe everything to the kind Marcelius, who tirelessly looked out for me. Also, I will always help gifted young men wherever this will be within my powers, this is a solemn oath I am taking. The only problem that would remain to be solved would be how to have our Lieserl with us; I wouldn't like for us to have to part with her. Ask your father, he is an experienced man and knows the world better than your impractical bookworm Johonzel. She shouldn't be stuffed with cow milk, this could make her stupid (yours must be much more nourishing, I believe, what do you think?!)

I got again a very self-evident but important scientific idea about molecular forces. You know that no noticeable evolution of heat takes place when two neutral liquids are mixed together. From this it follows, according to our theory of molecular forces, that there must exist an approximate proportionality between our constants Σc_α and the molecular volumes of the liquids. If this were true, then this would be the end of the molecular-kinetic theory of liquids. I'll see whether I can get hold of Ostwald or Landolt during the vacation period. I'll either stay here (for reasons of economy) or go to Zurich and work (this beats all secondary considerations).

My pupil has just told me that the thing with Bern is not so impossible after all. His mother seems to have become a little bit enlightened. But now this is not so important to me any longer, although it still would make me very glad.

Be cheerful, my dear, faithful little sweetheart, give my kind regards to your parents, and I clasp you tenderly to my heart. Your

Johonzel

128. TO MILEVA MARIĆ

Schaffhausen Tuesday [17 December 1901]

My dear sweetheart!

It is really a screamingly funny life that I am living here, completely in Schopenhauer's sense. That is to say, all day long I talk with nobody except my pupil. Even Mr. Baumer's company seems to me boring and insipid. I always find that I am in the best company when I am alone, except when I am with you. But you I miss very much,

I find that every regular guy must have a girl.

I would like so much to have you with me, even if you have a "funny figure," as you have written me twice already. Why don't you make a drawing of it for me, a really beautiful one! It will make me very happy if you make me a little pillow. But then you also must make the soul [inner slip cover?] for it (for the sake of the pillow), because I haven't the slightest idea where those of mine reside. You know the ghastly disorder that reigns among my worldly goods & it is really lucky that I don't have much. My new eating arrangement at the inn suits me quite well, in any case it is an enormous improvement, so that my moving to Bern, with my pupil, I mean, does not seem desirable any longer. But the people with whom I am eating here seem to me too stupid and commonplace. So I am sitting at the meals like a nutcracker & am playing with knife and fork between courses, while looking out of the window. The fellows must think that I have never laughed in my whole life; but, then, they have never seen me with my Doxerl.

I am now working very eagerly on an electrodynamics of moving bodies, which promises to become a capital paper. I wrote to you that I doubted the correctness of the ideas about relative motion. But my doubts were based solely on a simple mathematical error. Now I believe in it more than ever. Since that bore Kleiner hasn't answered yet, I am going to drop in on him [also: to take him to task; untranslatable pun] on Thursday. I want to induce him at all costs to let me work during the Christmas vacation. I'll see whether I'll succeed in that. To think of all the obstacles that these old philistines put in the way of a person who is not of their ilk, it's really ghastly! This sort instinctively considers every intelligent young person as a danger to his frail dignity, this is how it seems to me by now. But if he has the gall to reject my doctoral thesis, then I'll publish his rejection in cold print together with the thesis & he will have made a fool of himself. But if he accepts it, then we'll see what a position the fine Mr. Drude will take... a fine bunch, all of them. If Diogenes were to live today, he would look with his lantern for an *honest* person in vain.

I'll soon mail a small parcel for you and your sister, but nothing to eat, it's all reading stuff. Just don't you believe now that a lot is coming, it's not much but it's from the heart. But you must rejoice in it and think: "If he had more, he would send me more, my sweetheart." This is how you must think, right?!

Give my kind regards to your old lady & tell her also that I am looking forward to the thrashing she will bestow upon me one of these days.

Hearty kisses and hugs from your

Johonzel

I'll send you the money the moment you want it. It seems so funny when it comes from me. What do you think?

129. TO THE SWISS PATENT OFFICE

Schaffhausen, 18 December 1901

To the "Federal Office for Intellectual Property" [Patent Office]:

I, the undersigned, take the liberty of applying for the position of Engineer Class II at the Federal Office for Intellectual Property, which was advertised in the Bundesblatt [Federal Gazette] of 11 December 1901. I obtained my professional education in physics and electrical engineering at the School for Specialist Teachers of Mathematics and Physics at the Federal Polytechnikum in Zurich, which institution I attended from the Fall of 1896 to the Summer of 1900. There I obtained the Federal Diploma for Specialist Teachers after completion of my studies, based on an experimental project in physics and the successfully passed examination.

From the Fall of 1900 to the Spring of 1901 I lived in Zurich as a private teacher. At the same time I perfected my education in physics and wrote my first scientific paper. From 15 May to 15 July I was a substitute teacher of mathematics at the Technikum in Winterthur. Since 15 September 1901 I have been a tutor in Schaffhausen. During the first two months of my working here I wrote my doctoral dissertation on a topic in the kinetic theory of gases, which I submitted to Section II of the Faculty of Philosophy at Zurich University a month ago.

The documents that corroborate my statements are now at Zurich University, but I hope that I will be able to send them to you in a few days.

I am the son of German parents, but I have been living in Switzerland without interruption since age 16. I am a citizen of the City of Zurich.

Respectfully yours,

Albert Einstein
Bahnhofstr. Schaffhausen

130. TO MILEVA MARIĆ

Schaffhausen Thursday [19 December 1901]

My dear little sweetheart!

Good news once again! But stop, first my belated cordial congratulations on your birthday yesterday, which I forgot once again. But now listen & let me kiss you and hug you with joy! Haller has written me a friendly letter in his own hand, in which he requested that I apply for a newly created position in the patent office! Now there is no longer any doubt about it. Grossmann has already congratulated me. I am dedicating my doctoral thesis to him, to somehow express my gratitude. He has shown that he is a splendid fellow! And soon you'll become my happy little wife, you just watch and see. Our troubles have now come to an end. Only now that the terrible pressure of circumstances does not weigh upon me any longer do I really see how much I love you! I am sure that everything will be decided soon. Now I'll soon be able to clasp my Doxerl in my arms and to call her mine in front of the whole world. You'll soon be my "student" again, exactly as in Zurich. Are you glad?

Today I spent the whole afternoon with Kleiner in Zurich and explained my ideas on the electrodynamics of moving bodies to him & otherwise talked with him about all kinds of physical problems. He is not quite as stupid as I thought, and, moreover, he is a good guy. He said I may refer to him whenever I need a recommendation. Isn't that nice of him? He must go away during the vacation and he hasn't read the thesis yet. I told him to take his time, I am in no hurry. He advised me to publish my ideas about the electromagnetic theory of light of moving bodies together with the experimental method. He thought that the experimental method proposed by me is the simplest and most appropriate one conceivable. I was very pleased with the success. I shall certainly write the paper in the coming weeks. I am staying here during the vacation, but I'll spend the two feast days of Christmas with my sister in Paradies, in intimate winter solitude. If only you too could be there! But our Paradise will follow soon. I am absolutely delirious with delight. It is not yet certain whether the Englishman will come with me to Bern -- but under these circumstances I don't really care. The wretched old man will be surprised when I tell him about it! He is a miserable scoundrel. I've learned hair-raising things about him.

Hugs and kisses from your

Johonzel

The parcel with the books is already on its way.

131. TO MILEVA MARIĆ

[Schaffhausen] Saturday [28 December 1901]

My dearly beloved little sweetheart!

I am writing to you again, because I cannot bear not to write to you. What a precious sweetheart I have & what a fine little parcel has she sent me! She even hid a magnificently fine piece of tobacco in it & a most darling little letter. I rejoiced all day long. The goodies are incomparably good & I kiss you in my thoughts each time I eat one. I have eaten almost half of it, even though it arrived only yesterday at noon. I am decidedly not in a sad mood. All the time I rejoice in the fine prospects which are in store for us in the near future. Have I already told you how rich we will be in Bern? 3,500 fr. is the minimum salary the position pays according to the advertisement, but it increases up to 4,500. Ehrat thinks, though, that one cannot live on 4,000 fr. with a wife. But we will prove by deeds how fabulously that can be done! Isn't it so, sweetheart? After all, we managed in Zurich with barely half of that & had a great time of it. I find it quite funny how fussy people are. It is true that life is supposed to be more expensive in Bern than in Zurich. But this is probably not so serious.

Michele gave me a book on the theory of ether, written in 1885. One would think it came from antiquity, its views are so obsolete. It makes one see how fast knowledge develops nowadays. I now want to buckle down and study what Lorentz and Drude have written on the electrodynamics of moving bodies. Ehrat must get the literature for me. Grossmann is getting his doctorate on a topic that is connected with Fiedlering [fiddling: untranslatable pun] and non-Euclidean geometry. I don't know exactly what it is. Ehrat will do his thesis with Gaiser, as he refuses the ridiculous topic that Minkowski has proposed to him. You see, your Johonzel has been the first to finish his dissertation, even though he is a harassed little beast. When you are my dear little wife, we will zealously do scientific work together, so as not to become old philistines, right? My sister seemed to me so philistine. You must never become like that, it would be awful for me. You must always remain my witch and my street urchin. I long very much after you. If only I could have you just a little bit! Apart from you, all the people look so alien to me as if they were separated from me by an invisible wall. You should just hear Ehrat talking about marriage, it's too funny. He talks about it as about a bitter medicine that has simply to be taken dutifully. His wife will find it amusing. How differently people look at one and the same thing, it's very funny, isn't it.

Tender kisses from your

Johonzel

132. RECEIPT FOR THE RETURN OF DOCTORAL FEES.

Zurich, 1 February 1902

No. 124.

Receipt

The undersigned herewith confirms that the doctoral fees of 230 (two hundred thirty) fr., paid on 23 Novbr. 1901, have been returned to him today in cash by the University Office with the concurrence of the Dean, Prof. Dr. Hans Schinz.

Albert Einstein

133. TO CONRAD HABICHT

[Bern, 4 February 1902]

Dear Habicht!

I sailed away from N[üesch] with spectacular effect. Come and see me when you have time, I'll tell you a good story. I would have visited you already, but it is so far & the chances of finding you at home are not great.

With kind reg. your

Albert Einstein
Gerechtigkeitsgasse 32
1st floor

134. TO MILEVA MARIĆ

Bern Tuesday [4 February 1902]

My beloved sweetheart!

Poor, dear sweetheart, you must suffer enormously if you cannot even write to me yourself! And our dear Lieserl too must get to know the world from this aspect right from the beginning! I hope that you will be up and around again by the time my letter arrives. I was scared out of my wits when I got your father's letter, because I had already suspected some trouble. External fates are nothing compared with this. At once I felt like being a tutor with N[üesch] for two more years if this could make you healthy and happy. But you see, it has really turned out to be a Lieserl, as you wished. Is she healthy and does she already cry properly? What kind of little eyes does she have? Whom of us two does she resemble more? Who is giving her milk? Is she hungry? And so she is completely bald. I love her so much & I don't even know her yet! Couldn't she be photographed once you are totally healthy again? Will she soon be able to turn her eyes toward something? Now you can make observations. I would like once to produce a Lieserl myself, it must be so interesting! She certainly can cry already, but to laugh she'll learn much later. Therein lies a profound truth. When you feel a little better, you must make a drawing of her.

It's delightful here in Bern. An ancient, exquisitely cozy city, in which one can live exactly as in Zurich. Very old arcades stretch along both sides of the streets, so that one can go from one end of the city to the other in the worst rain without getting noticeably wet. The homes are uncommonly clean, I saw this everywhere yesterday when I was looking for a room. It does me extremely good to have escaped from the unpleasant environment. I already saw to it that an advertisement will be published in the local gazette. I hope it will be of some help. If only I got 2 lessons per day, I could save something for you. I have a large beautiful room with a very comfortable sofa. It only costs 23 fr. This is not much, after all. In addition, 6 upholstered chairs and 3 wardrobes. One could hold a meeting in it. Its plan follows:

[Plan]

B =	little bed	N =	chamber pot & table
b =	little picture	F =	little window
d =	coverlet	O =	stove
gS =	magnificent chair	S =	little chair
$g\Sigma$ =	magnificent mirror	T =	door
J =	Johonzel	τ =	table
K =	wardrobe	ν =	nothing
κ =	sofa	U =	little clock
$k\Sigma$ =	small mirror	Γ =	look at that!

[sketch]

And now, get well again fast, I beg you as nicely as I can. Give my kindest regards to your dear mother.

And for you, tender kisses from your loving

Johonzel
Gerechtigkeitsgasse 32
Bern
(c/o Mrs. Sievers)

135. ADVERTISEMENT FOR PRIVATE LESSONS

[5 February 1902]

Private lessons in

MATHEMATICS AND PHYSICS

for students and pupils

given most thoroughly by

ALBERT EINSTEIN, holder of the fed.

polyt. teacher's diploma

GERECHTIGKEITSGASSE 32, 1ST FLOOR.

Trial lessons free.

136. TO MILEVA MARIĆ

Bern Saturday [8? February 1902]

My dear little sweetheart!

Guess where I was today! At a class of forensic <me> pathology with demonstration ad oculos. Friend Frösch, who also happens to be here, took me along. The thing interested me so tremendously that from now on I'll go there every Saturday. A sixty-year-old woman was questioned, who tried to commit arson while blind drunk, as well as a man accused of swindling, who seems to suffer from megalomania. (There are interesting cases of pathological swindlers in Forel.) Because of his great intelligence, Frösch enjoys such respect on the part of the professor, that he turns to Frösch every time he made some smart remark. I then spent the rest of the afternoon with Frösch.

The situation with the private lessons isn't bad at all. I have already found two gentlemen, an engineer & an architect & more in prospect. I am going to give both of them together a kind of a private course & am getting 2 fr. per man and lesson. This is quite nice, isn't it. It is going to start on the evening of the day after tomorrow.

I am now expounding to Habicht the paper I have handed in to Kleiner. He is quite enthusiastic about my good ideas & nags me to send to Boltzmann the part of the paper that refers to his book. I am going to do it.

I have almost finished reading Mach's book with tremendous interest, and will this evening

137. TO MILEVA MARIĆ

[Bern] Monday [17? February 1902]

My little sweetheart!

In defiance of all my pedagogical principles, I am writing you again today, because I got such a nice little letter from you. You needn't be jealous of Habicht and Frösch, because what are they to me compared with you! I long after you every day, but I do not act that way because it's not "manly" and I always think: Joggele, you go first, you are the man after all. But it's true nevertheless that it is very nice here. But I would rather be with you in some backwater than without you in Bern, that's for sure. And no pleasure I am having here compares with that of getting a little letter from the little sweetheart. Without you even studying gives me only half the pleasure. Recently a man who used to be at the Polytechnikum in Zurich and is now employed at the patent office started a conversation with me. He finds that it's very boring there -- certain people find everything boring -- I am sure that I will find it nice and that I will be grateful to Haller as long as I live. He also told me that the advertisement in the Federal Gazette under the announcements of vacant positions does not otherwise mention "physics", but only polytechnical-mechanical education. Haller inserted this for my sake. Further, the election of employees happens in such a way that Haller makes the proposal and the Federal Council appoints; hence, there is hardly any doubt. Besides, he added, a little contemptuously, the position is of the lowest rank and it is unlikely that anyone will compete with me for it. I was very glad to hear that. We two don't give a damn about "height"!

138. PAULINE EINSTEIN TO PAULINE WINTELER

[Milan] Thursday, 20 February 1902

Dear Frau Professor!

[...]

Maja hasn't written a word about the event, knowing well that in this case she wouldn't have found a sympathetic listener in me either. She knows that we strongly oppose the liaison of Albert and Miss Marić, that we don't want ever to have anything to do with her & that there is constant friction with Albert because of it. For this reason alone, she should not talk anywhere about an engagement of her brother, least of all in Aarau, where also the relationship with your house has to be considered. We had a violent argument with our daughter here in Mettmenstetten about this, so that we bade a very frosty goodbye to each other.

This Miss Marić is causing me the bitterest hours of my life, if it were in my power, I would make every possible effort to banish her from our horizon, I really dislike her. But I have lost every influence on Albert. You can imagine, d[ear] Fr[au] Professor, how unhappy this makes me!

[...]

Pauline Einstein

"ON THE THERMODYNAMIC THEORY OF THE DIFFERENCE IN POTENTIALS BETWEEN METALS AND FULLY DISSOCIATED SOLUTIONS OF THEIR SALTS AND ON AN ELECTRICAL METHOD FOR INVESTIGATING MOLECULAR FORCES."

April? 1902

(Text in Vol. 2)

139. TO CONRAD HABICHT

[Bern, April? 1902]

Dear Habicht!

Ehrat whines to me per postcard that he cannot get "Archimedes opera" back from you. He must hand it in on Monday without fail, otherwise his lord and master R[udio] will growl at him in High German. So on to the Post Office!

Until a pleasurable r[eunion], your

E

"KINETIC THEORY OF THERMAL EQUILIBRIUM AND OF THE SECOND LAW OF THERMODYNAMICS."

June 1902

(Text in Vol. 2)

140. THE SWISS DEPARTMENT OF JUSTICE TO THE SWISS FEDERAL COUNCIL

[Bern, 2 June 1902]

THE DEPARTMENT OF JUSTICE & POLICE
OF THE SWISS CONFEDERATION

THE FEDERAL OFFICE for INTELLECTUAL PROPERTY

To the Federal Council

In re:
Election of two engineers
for the Office for Int. Property.

[...]

According to the view and experience of the Director of the Office for Intellectual Property, in general, preparation at a technikum is insufficient for a fruitful activity as a technical expert of the Office. The remaining candidates were examined by the Director partly in writing and partly orally. The examination of the gentlemen named under 1, 2, and 4 below turned out in favor of Messrs. Schenk and Einstein.

[...]

We therefore propose the election of Messrs. Schenk and Einstein. Both could assume their positions on 16 June at the latest.

It is recommended, as has been done heretofore, to make at first only provisional appointments, subject to later definitive confirmation and the establishment of a salary commensurate with the performance at that time.

Thus, it is moved that the Federal Council resolve the following:

1.) Messrs. J. Heinrich *Schenk* from Röthenbach (Bern) and Albert *Einstein* from Zurich will be provisionally elected as Technical Experts Class 3 <(Engineers Class 2)> of the Fed. Office for Intellectual Property at an annual salary of 3,500 fr. each.

2.) It is left to the Director of the Office to decide on the date on which the two men elected will assume their duties.

[...]

Federal
Department of Justice & Police:
Brenner

Attached:
Application file.

141. FROM THE SWISS DEPARTMENT OF JUSTICE

Bern, 19 June 1902

Department of Justice and Police
of the Swiss Confederation

Fed. Office for Intellectual Property

Mr. Albert Einstein, Bern

Highly esteemed Sir!

At its session of 16 June 1902, the Federal Council elected you provisionally as a Technical Expert Class 3 of the Fed. Office for Intellectual Property with an annual salary of 3,500 fr.

Respectfully.

Federal
Department of Justice & Police:
Brenner

142. FROM THE SWISS PATENT OFFICE

[Bern, 19 June 1902]

Mr. J. Heinrich Schenk, Engineer
Zurich I, Mühlegasse 31, 3d [floor]

Mr. Albert Einstein
Bern, Thunstr. 43a

Highly esteemed Sir!

Herewith we transmit to you a letter of appointment of the Fed. Department of Justice & Police, and expect that you will assume your duties on this coming 1 July at the latest.

However, you can also start earlier.

Respectfully

O[ffice] f[or] I[ntellectual P[roperty]